W9-CHL-488

Case Studies for Understanding the Human Body
Second Edition

Stanton Braude

Washington University in St. Louis and
The International Center for Tropical Ecology

Deena Goran

John Burroughs School and
the University College of Washington University

Alexander Miceli

Washington University in St. Louis

JONES & BARTLETT
LEARNING

World Headquarters

Jones & Bartlett Learning
40 Tall Pine Drive
Sudbury, MA 01776
978-443-5000
info@jblearning.com
www.jblearning.com

Jones & Bartlett Learning Canada
6339 Ormindale Way
Mississauga, Ontario L5V 1J2
Canada

Jones & Bartlett Learning International
Barb House, Barb Mews
London W6 7PA
United Kingdom

Jones & Bartlett Learning books and products are available through most bookstores and online booksellers. To contact Jones & Bartlett Learning directly, call 800-832-0034, fax 978-443-8000, or visit our website, www.jblearning.com.

Substantial discounts on bulk quantities of Jones & Bartlett Learning publications are available to corporations, professional associations, and other qualified organizations. For details and specific discount information, contact the special sales department at Jones & Bartlett Learning via the above contact information or send an email to specialsales@jblearning.com.

Production Credits

Publisher, Higher Education: Cathleen Sether
Acquisitions Editor: Molly Steinbach
Senior Associate Editor: Megan R. Turner
Editorial Assistant: Rachel Isaacs
Associate Production Editor: Jill Morton
Senior Marketing Manager: Andrea DeFronzo
V.P., Manufacturing and Inventory Control: Therese Connell
Composition: Circle Graphics
Cover Design: Scott Moden
Assistant Photo Researcher: Rebecca Ritter
Cover Image: © YenLev/ShutterStock, Inc.
Printing and Binding: Courier Corporation
Cover Printing: Courier Corporation

Library of Congress Cataloging-in-Publication Data

Braude, Stanton, 1961- author.
  Case studies for understanding the human body / Stanton Braude, Deena Goran, Alexander Miceli.—Second edition.
    p. ; cm.
  ISBN: 978-1-4496-0499-8 (alk. paper)
 1. Human biology—Case studies. I. Goran, Deena, author. II. Miceli, Alexander, author.
III. Chiras, Daniel D. Human biology. c2011. Related to (work): IV. Title.
  [DNLM: 1. Physiological Phenomena—Case Reports. 2. Physiological Phenomena—Problems and Exercises. 3. Anatomy—Case Reports. 4. Anatomy—Problems and Exercises. 5. Disease—Case Reports. 6. Disease—Problems and Exercises. QT 18.2]
  QP36.B858 2012
  612—dc22

                            2010047783

6048

Printed in the United States of America
15 14 13 12 11   10 9 8 7 6 5 4 3 2 1

Contents

Preface

Human anatomy and physiology (A&P) and human biology courses are typically offered for students with widely varying degrees of motivation and background. Students range from nursing students, who may previously have taken only high school biology, to undergraduate biology majors, who may already have taken several advanced courses. Because human A&P is an extremely content-rich course, it is typically taught in a frontal lecture format. Reading the text may be the only student-directed component of the course. We have written a series of case study exercises for the diverse audience of students to enrich their learning experiences both in and out of the classroom.

We have used these case study exercises successfully in several of our own courses for various audiences: non–science major students enrolled in our human biology course; nursing students enrolled in our A&P course; and undergraduate biology majors enrolled in our A&P course. In each situation, we believe that these case study exercises were extremely successful at facilitating better comprehension of complex physiological concepts. We use these exercises in a cooperative learning setting, where students work together in small groups to review relevant material and then solve the open-ended questions about the case. These exercises may also be used very effectively as homework problems for students to work on individually.

We have expanded this *Second Edition* in many different ways to improve upon the foundation set forth by the *First Edition*. We have broadened the range of the case studies to include examples from literature, films, and pop culture. We hope that these new narratives will intrigue students. Our goal for all the case studies, but especially the newer additions, is to convey human physiology in a contextual manner that makes the intricate biological details more tangible for the students. By thinking about cardiac function through the eyes of Indiana Jones or by studying the lymphatic system from the point of view of the Elephant Man, we hope to provide a situation where students are more engaged to learn about human physiology. We also hope to give educators a venue for energizing their students with these engaging cases.

Each of our exercises begins with a short narrative about the patient or the particular subject matter. The narrative is deliberately written in casual language to draw the reader into the case. The patients of these cases range from an athlete injured during a soccer game to a family member who experiences abdominal pain during Thanksgiving dinner. The questions that follow are designed to lead the student through an understanding of the case. At first, these questions ask the students to describe and review specific physiological processes that will be necessary to better understand the case as a whole. For example, the questions about the actual case may lead the students through a calculation or a line of reasoning necessary to solve the final questions about a diagnosis or treatment. Many of the cases end with "Integrative Challenge Questions" aimed at pointing out the interactions between different systems. All cases use the narrative to absorb the student in solving a problem based on understanding the underlying anatomy and physiology.

The series of cases in our text covers common diseases of all of the major organ systems. In addition, we include exercises on other related topics that are often part of human A&P or human biology texts. These range from the genetics of fragile X syndrome to the evolution of drug resistance in bacterial populations. The exercises are broad enough in scope to be used in any A&P course, but we realize that every one of these exercises might not be appropriate for every course. As with any text, we anticipate that the instructor will assign the subset of chapters that works best for his or her course. We have provided a surplus of cases so that instructors should be able to find ones that fit his or her syllabus and his or her students.

Although we appreciate the value of laboratory experience in teaching science, our text is not a collection of laboratory exercises, but rather it is a collection of thinking exercises. No additional materials or resources are necessary to use these exercises in a class. Just as lab exercises may teach some of the techniques that science majors will use later in their careers, our exercises develop communication and critical thinking skills that *all* students will use throughout their lives.

About the Authors

Stanton Braude (BS, Biology; MS, Biology; PhD, Ecology and Evolutionary Biology) is on the faculties of Washington University and the University of Missouri in St. Louis. He has been teaching biology since 1983 and was recognized as the Outstanding College Biology Teacher of the Year by the National Association of Biology Teachers in 2004.

Deena Goran (BA, Biology; BS, Physical Therapy; MA, Education) is on the faculty of the John Borroughs School and the University College of Washington University. She has taught in the physical therapy program at Washington University School of Medicine; has taught anatomy and physiology at University College; and teaches middle and upper school biology at John Borroughs.

Alexander Miceli (BA, Chemistry and Italian; MA, Biology) is a candidate for a PhD in molecular cell biology at Washington University in St. Louis. He is a regular lecturer in anatomy and physiology and human biology.

Acknowledgments

Special thanks to Holly Epple, Shirly Mildiner, Ritesh Agrawal, and Leah Corey, who field tested many of these exercises. Thanks also to Nancy Berg, PhD, David Goran, MD, Karen Norberg, MD, and Elizabeth Hansen, MD, for their help with details, large and small.

1. Hydration and Dehydration (Water and Osmosis)

The Berg family went to the community pool to beat the August heat. Everyone was playing hard—swimming, diving, and jumping. After a few hours, the children, Ellie and David, noticed that their fingers and toes were getting wrinkled, just like what happens when they stay in the bath for too long. They wondered why this happens.

Ellie's father is a chemical engineer, and he told Ellie that water and dissolved particles were drawn out of her skin by diffusion into the large body of water around her. Her mother is a chemist, and she told Ellie that water was actually drawn into her skin by osmosis. Let's help Ellie and David figure out who is right.

1. The osmolarity of a solution is the sum of the concentrations of all the different dissolved particles. If a liter of water has one mole of glucose, one mole of sodium, and one mole of chloride dissolved in it, what is the osmolarity of the solution?

2. Solutions of lower osmolarity are hypotonic, and solutions of higher osmolarity are hypertonic. Solutions with the same concentration of dissolved particles (even if they are different kinds of particles) are isotonic. Water tends to move from areas of lower osmolarity into areas of higher osmolarity.

   Physiological fluids have an osmolarity of approximately 300 mOsm/L. In contrast, tap water has far fewer dissolved particles.

   a. Is clean bath water likely to be hypertonic or hypotonic compared to Ellie's tissues?

b. In which direction will the water move (into or out of Ellie's tissues)?

3. Chlorine is added to swimming pool water to inhibit the growth of bacteria, algae, and other microbes. At very low levels, approximately 2 mg/L of water, chlorine in a large body of water, like a swimming pool, has no toxic effects on human cells. How does the addition of chlorine to the pool water affect its osmolarity?

4. Do you think the addition of chlorine has changed the water tonicity relative to David's tissues? Explain.

5. Epithelial tissue, which constitutes the epidermal layer of skin, is made up of cells attached to a basement membrane. Because the cells are tightly packed and anchored to the basement membrane, their swelling causes contortions of the basement membrane and also of the cells themselves (kind of like the wrinkles in wet carpeting). Ellie's father tells her that his fingers used to wrinkle up when he went swimming in Lake Michigan (a fresh water lake) as a little boy. Ellie's mother tells her that never happened when she went swimming at the ocean in Maine when she was a little girl. Explain why.

6. What would you expect to happen to the fingers and toes of swimmers in the Great Salt Lake in Utah? Would their skin wrinkle or feel tight? Explain why.

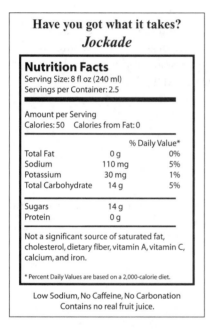

Purest Spring Water

Nutrition Facts

Serving Size: 8 fl oz (240 ml)

Servings: about 4

Calories: 0

Amount per Serving		% Daily Values*
Calories: 0		
Total Fat: 0g		0%
Sodium: less than 5 mg		0%
Total Carbohydrate: 0g		0%
Protein: 0g		

* Percent Daily Values are based on a 2,000-calorie diet.

Mineral Composition in mg/liter

Calcium	78	Bicarbonates	357
Magnesium	24	Sulphates	10
Silica	14	Chlorides	4
Nitrate (N)	1		

Neutrally balanced pH = 7.2

Figure 1.1. Spring water label.

Have you got what it takes?
Jockade

Nutrition Facts

Serving Size: 8 fl oz (240 ml)
Servings per Container: 2.5

Amount per Serving
Calories: 50   Calories from Fat: 0

		% Daily Value*
Total Fat	0 g	0%
Sodium	110 mg	5%
Potassium	30 mg	1%
Total Carbohydrate	14 g	5%
Sugars	14 g	
Protein	0 g	

Not a significant source of saturated fat, cholesterol, dietary fiber, vitamin A, vitamin C, calcium, and iron.

* Percent Daily Values are based on a 2,000-calorie diet.

Low Sodium, No Caffeine, No Carbonation
Contains no real fruit juice.

Figure 1.2. Sports drink label.

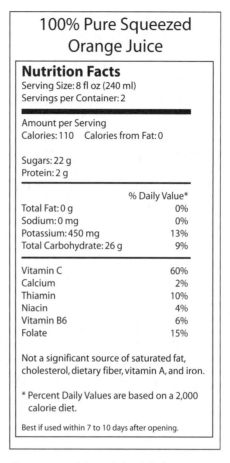

100% Pure Squeezed Orange Juice

Nutrition Facts

Serving Size: 8 fl oz (240 ml)
Servings per Container: 2

Amount per Serving
Calories: 110   Calories from Fat: 0

Sugars: 22 g
Protein: 2 g

	% Daily Value*
Total Fat: 0 g	0%
Sodium: 0 mg	0%
Potassium: 450 mg	13%
Total Carbohydrate: 26 g	9%
Vitamin C	60%
Calcium	2%
Thiamin	10%
Niacin	4%
Vitamin B6	6%
Folate	15%

Not a significant source of saturated fat, cholesterol, dietary fiber, vitamin A, and iron.

* Percent Daily Values are based on a 2,000 calorie diet.

Best if used within 7 to 10 days after opening.

Figure 1.3. Orange juice label.

Pops Super Diet Cola

16.9 FL OZ
(1.06 PT)
500 ml

Please Recycle

Nutrition Facts

Serving Size: 8 fl oz (240 ml)

Servings: 2

Amount per Serving	% Daily Values*
Total Fat: 0 g	0%
Sodium: 30 mg	1%
Total Carbohydrate: 0 g	0%
Protein: 0 g	

* Percent Daily Values are based on a 2,000-calorie diet.

Figure 1.4. Diet soda label.

7. After four hours of playing, everyone was ready to slow down, cool off, and walk home for dinner. Ellie challenged her brother to a race home. After their one-mile run, both were soaked with sweat. David let them into the house and opened the refrigerator to get a drink. They found diet soda, orange juice, bottled water, and a sports drink (Figures 1.1, 1.2, 1.3, and 1.4). Which should they drink and why? Examine the contents of each product and discuss the effects they might have on David's and Ellie's dehydration.

INTEGRATIVE CHALLENGE QUESTIONS

8. Explain what would happen to blood flow and blood pressure if David and Ellie were running in cold weather.

9. In cold weather, peripheral blood vessels are constricted and blood is shunted into the body core. This raises blood pressure to the organs. How would the increased blood pressure, especially in the large vessels supplying the brain and heart, affect the secretion of antidiuretic hormone (ADH)?

10. How would the increased blood pressure affect the secretion of atrial natriuretic peptide?

11. How would these two hormones then affect urine output and blood volume? Based on your answers, can cold weather also cause dehydration?

(Read more about this system in Daniel Chiras, *Human Biology, 7th edition*, Chapter 3.)

2. Nurses Houlihan and Ratched (Osmosis and Diffusion)

→ *Ratched doesn't know basic bio or chem.*

Margaret Houlihan and Mildred Ratched were roommates during nursing school. They are both committed to becoming qualified, practicing nurses. However, Ratched hated the basic science classes and wanted to move on to the clinical work. Houlihan has been a good friend and has helped Ratched pass her basic chemistry and biology courses. This semester they are finally getting to learn some practical skills.

Last week, Houlihan and Ratched learned about intramuscular (i.m.) and intravenous (i.v.) injections. Their instructor suggested that they get the feel for the syringe by injecting water into an orange. Ratched did that last night and was eager to practice for real. She decided to practice on herself, but she knew it was dangerous to inject anything that was not sterile. She brought home a small bottle of distilled water with which to practice. Ratched gave herself three i.m. injections in the arm and felt pretty sore. She was confused because she used a small-gauge needle and small volumes of water. Just as she was about to practice an i.v. injection, Houlihan came in and gasped. She was shocked at what her roommate had done and grabbed the syringe away from her before she could do worse damage. Ratched didn't understand why her arm was sore or why Houlihan reacted so severely. Houlihan knew Ratched really hadn't paid attention when they were supposed to be learning about osmosis and diffusion. Help her explain to her roommate why we need to practice injections with a physiologically isotonic solution.

1. Was Ratched injecting herself with a hypotonic or hypertonic solution? Explain.

2. What happens to cells when they are in a hypertonic solution?

    a. What happens to cells when they are in a hypotonic solution?

    b. What happens to cells when they are in an isotonic solution?

    c. What concentration of salt is used in a physiologically isotonic solution?

3. Water can enter a cell through several mechanisms.

    a. Explain the differences between facilitated diffusion and active transport.

    b. What role do carrier proteins have in these processes?

4. When Ratched injected distilled water into her muscle, where did that water go?

a. What happened to the cells surrounding the point of injection?

5. What would happen if she injected distilled water directly into a vein?

6. Houlihan graduated with honors and served in the 4077th M.A.S.H. unit in South Korea.

    a. Which i.v. solution would she have used to rehydrate unconscious casualties?

    b. Why wouldn't she use distilled water?

INTEGRATIVE CHALLENGE QUESTIONS

7. Mildred Ratched worked in the Salem Oregon Mental Institute until she was attacked by Randle McMurphy, a patient who refused to submit to her authority. She finally found work at the Portland Regional Dialysis Center, supervising the giant dialysis machines that help patients whose kidneys no longer function properly. Briefly describe how the kidneys separate urine from blood.

8. The kidneys perform their tasks by several processes. Give an example of a substance that is "handled" by the kidneys through:

   a. Simple diffusion:

   b. Active transport:

   c. Secretion:

9. Which substances move by diffusion in one part of the nephron and then move by active transport in another section of the nephron?

10. What would happen if Nurse Ratched decided she didn't like the attitude of one of her dialysis patients and replaced the dialysis solution with distilled water?

11. How would Nurse Ratched sedate an agitated patient during his dialysis treatment?

(Read more about this system in Daniel Chiras, *Human Biology, 7th edition*, Chapter 3.)

3. J. Wellington Wimpy (Basic Biochemistry)

J. Wellington Wimpy has struggled with weight and finances his whole life. Finally, his friends coordinated an intervention and refused to continue loaning him funds to feed his burger addiction. Wimpy was forced to change his ways and adopted a healthier, turkey and veggie sandwich diet. Trading overpriced burgers for affordable sub sandwiches has improved his health and repaired his relationships with his friends. Wimpy even jogs with Olive a few mornings a week.

Let's take a look at what's in Wimpy's favorite sandwiches.

1. Wimpy's 6-inch turkey sub contains 47 grams of carbohydrate.

   a. What part of the sandwich contains the complex carbohydrate, starch?

   b. When we digest the starch, which monosaccharides are released?

   c. Of the 47 grams of carbohydrate in the turkey sub, 6 grams are simple sugars. When we digest the disaccharide sucrose, which monosaccharides are released?

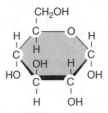

Figure 3.1. Glucose.

d. How does your body use these carbohydrates (Figure 3.1)?

2. Wimpy's 6-inch turkey sub also contains 5 grams of dietary fiber.

   a. Which kind of carbohydrate is this fiber?

   b. Even though we cannot digest it, how do we benefit from fiber in our diet?

   c. Give some examples of "simple carbohydrates" and "complex carbohydrates." Name some examples of common monosaccharides and disaccharides.

   d. Both types of carbohydrates are used as glucose sources. Are there any differences in how these two types of carbohydrates deliver their "glucose" to the body?

e. Which type makes glucose available more quickly?

f. Which type makes glucose available over a longer time period?

g. How are glucose polymers stored in the body?

h. Is there a similar mechanism of storing glucose in plants?

3. If Wimpy drinks a diet cola with his sandwich, he is not adding to the total caloric content of the meal. What can we infer about the metabolism of artificial sweeteners in Wimpy's digestive tract?

4. Wimpy's turkey sub contains 2.5 grams of fat.

   a. Fat is one kind of lipid. How many fatty acids are bound to a backbone of glycerol in each molecule of fat?

b. Which kind of lipid is the major component of cell membranes?

c. How many fatty acids and how many phosphates are bound to the backbone of glycerol in each of these lipid molecules?

d. What other important jobs do fats and lipids do in your body?

5. Explain the differences between saturated fatty acids and nonsaturated fatty acids. Provide examples of each. Identify the saturated, monounsaturated, and polyunsaturated fatty acids in this fat (Figure 3.2).

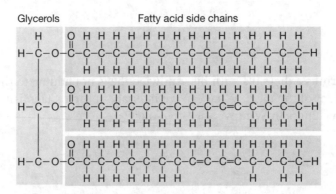

Figure 3.2. Triglyceride.

6. Both Wimpy's turkey sub and his veggie delight have 0 grams of cholesterol. What is the benefit of a low-cholesterol diet?

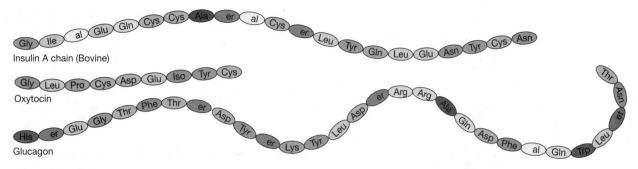

Figure 3.3. Examples of proteins.

7. Wimpy's turkey sub has 18 grams of protein and his veggie sub has 17 grams of protein.

    a. List at least 6 different jobs done by proteins (Figure 3.3) in your body.

    b. What are the molecular building blocks of proteins?

8. Nucleic acids are not listed on nutrition labels, but they are essential biochemicals in our cells.

    a. What do nucleic acids (Figure 3.4) do in our cells?

    b. Nucleotides are the building blocks of nucleic acids. Name the five different nucleotides.

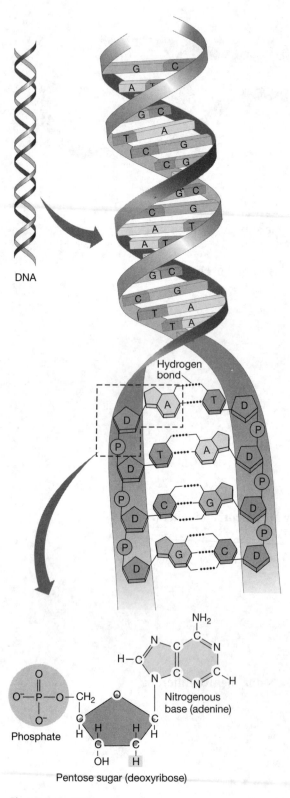

DNA

Hydrogen bond

NH₂

Phosphate

Nitrogenous base (adenine)

Pentose sugar (deoxyribose)

Figure 3.4. DNA.

c. Which nucleic acid is involved in energy transfer inside your cells?

9. Wimpy's subs also contain vitamin A, vitamin C, iron, and calcium. Explain how our bodies use each of these vitamins and minerals.

(Read more about this system in Daniel Chiras, *Human Biology*, *7th edition*, Chapters 3 and 5.)

4. Lactose Intolerance (Enzymes)

Carol was recovering from a terrible bout with the flu, having been sick for nearly two weeks. As she was getting her strength back, her appetite also improved. In fact, after not eating very much for almost two weeks, Carol was famished. But something strange happened to her: When she began eating milk products after the illness, she experienced terrible abdominal distress—she had bloating, pain, and diarrhea.

Carol went to see her doctor and explained her symptoms. Her doctor said that the virus she contracted must have affected the enzyme-producing cells in her small intestine. Carol had become lactose intolerant. Whenever she ate any food containing milk, her digestive system was unable to break down the sugar in the food or beverage. Her doctor suggested that Carol stay away from dairy products, except for yogurt.

Last night Carol saw you studying your biology textbook in the library, and she asked you to explain some things to her.

1. Carol's first question was, "What is an enzyme?"

    a. Use the enzyme lactase as an example, and explain how it functions (Figure 4.1).

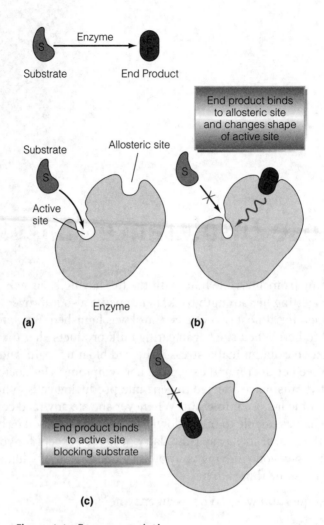

Figure 4.1. Enzyme regulation.

b. What is the substrate for lactase?

c. What are the end products of this reaction?

2. Carol now has some idea what you meant by protein but wants to know what you meant by "catalyst." Explain.

3. Carol partly understood that, but what exactly did you mean by "activation energy"?

4. You told Carol that there are thousands of different enzymes in her body.

   a. Why can't one of the other ones break down lactose?

   b. Why are they all so specific?

5. Why would Carol's doctor suggest that she eat yogurt instead of other milk products?

INTEGRATIVE CHALLENGE QUESTIONS

6. Why would the lack of lactase cause Carol so much distress? Where would the undigested lactose travel from the small intestine?

    a. What could cause the production of gas from the breakdown of sugar?

7. Provide some examples of other enzymes in the body.

    a. What other functions can enzymes serve?

    b. Are they important only for digestion, as is the case with lactase?

8. What is an allosteric site on an enzyme?

    a. What function does it serve?

    b. Why would it be important to regulate the function of an enzyme at all?

(Read more about this system in Daniel Chiras, *Human Biology, 7th edition,* Chapter 3.)

5. Animal House (Cells and Organelles)

Larry and Kent are freshmen at Faber University, and they are eager to join a fraternity. Luckily for them, Delta House welcomes everyone who has more important things to do than study.

As soon as they entered Delta House, Larry and Kent were handed a couple of cold beers. Larry said he really didn't like the taste but he didn't want to pass judgment too quickly. The guys tried to keep up with their new friends, and to appreciate the full flavor of the brewed malt and hops. But after Bluto topped them up with yet another round, the boys began to feel sick. Larry made it to the porcelain idol, but Kent tossed his cookies right in the middle of the kitchen. This didn't seem to faze their host, or anyone else at Delta House, and the party continued all around them.

In their livers, cells that produce the enzymes alcohol dehydrogenase and aldehyde dehydrogenase were busy at work. Alcohol dehydrogenase detoxifies ethanol by converting it to acetaldehyde. A buildup of acetaldehyde triggers nausea and vomiting, but the enzyme aldehyde dehydrogenase can convert acetaldehyde into a less toxic form, acetic acid. For some reason, the older residents of Delta House didn't seem to be suffering the consequences of acetaldehyde buildup,

1. How do liver cells make an enzyme like alcohol dehydrogenase or aldehyde dehydrogenase (Figure 5.1)?

    a. Describe the steps involved in the expression of the alcohol dehydrogenase gene.

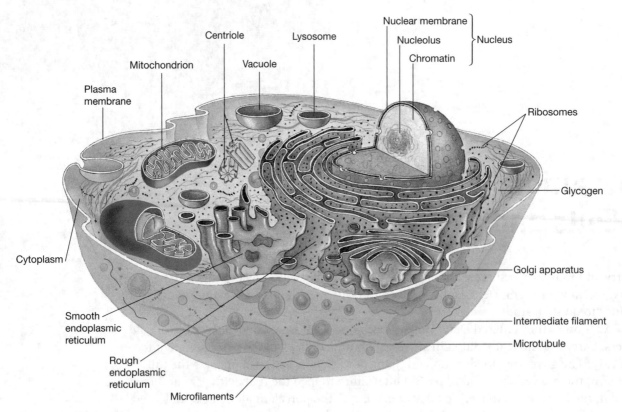

Plasma membrane

Mitochondrion

Centriole

Vacuole

Lysosome

Nuclear membrane

Nucleolus — Nucleus

Chromatin

Ribosomes

Glycogen

Cytoplasm

Golgi apparatus

Smooth endoplasmic reticulum

Intermediate filament

Microtubule

Rough endoplasmic reticulum

Microfilaments

Figure 5.1. The cell.

b. What role do ribosomes play in the process?

2. Despite his generous girth, Bluto's diet seems to consist of little more than alcohol. The food piled on his tray in the cafeteria is just ammo for a food fight. But Bluto could get plenty of energy from the alcohol he drinks. Which organelles in Bluto's liver use the NADH that is generated during detoxification of alcohol (Figure 5.2)?

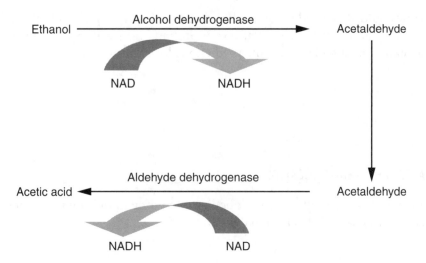

Figure 5.2. Alcohol metabolism.

3. Eventually cells will degrade defective organelles to maintain optimal cellular integrity.

   a. What is the specialized organelle to which those defective organelles are sent for destruction?

   b. What does this specialized organelle contain that breaks down defective materials?

4. Bluto doesn't think his tolerance of alcohol has anything to do with enzymes. He insists that the alcohol goes into his stomach, not his cells. But Larry and Kent have been going to class occasionally. Kent explains that molecules are absorbed into the bloodstream from the submucosa of the digestive tract, but he can't remember how molecules get into cells. Help him out. First describe the plasma membrane of his liver cells. How do molecules cross this membrane?

5. How could we test whether alcohol gets into the liver cell by simple diffusion, or whether it requires a special carrier protein in the cell membrane? Suggest an experiment and explain the predictions for each mechanism.

INTEGRATIVE CHALLENGE QUESTION

6. Alcohol has other effects on our bodies, including inhibiting the hormone ADH. How would this lead to dehydration?

(Read more about this system in Daniel Chiras, *Human Biology, 7th edition*, Chapter 3.)

6. The Elephant Man (Lymphatic System)

With his whole heart and soul the Elephant Man believed his condition was the result of a tragic accident involving his mother. . . . The mother-to-be found herself pushed into the road, under the feet of an elephant. She was nearly frightened to death. Some time after her baby was born he began to show the marks of her terrifying experience. He developed an arm as big as an elephant's leg. A head domed like an elephant's. And later this warty, elephant-like skin.

[His] body was covered with loose warty skin that could be drawn up in immense folds. One flap reached from the buttocks down to the middle of the thigh . . . The big right arm was three times the size of the left. The wrist was an incredible twelve inches around. The biggest of the fingers of the giant right hand measured five inches around.

—Frederick Drimmer, *The Elephant Man,* 1985,
G. P. Putnam's Sons, New York

Joseph Merrick lived in England at the turn of the nineteenth century. He was commonly referred to as the "Elephant Man" based on his appearance. After working in side shows in London, Merrick came in contact with the physician Frederick Treves. Several depictions of their relationship and of Merrick's life have been made over the years. Some believe that Merrick's condition resulted from elephantiasis, which involves blockage of the lymphatic vessels. Let us consider how this condition might have resulted in symptoms described previously.

1. Lymphatic vessels carry extra cellular fluid back to the bloodstream. How are lymph vessels and veins similar in the way they move fluid in one direction (Figure 6.1)?

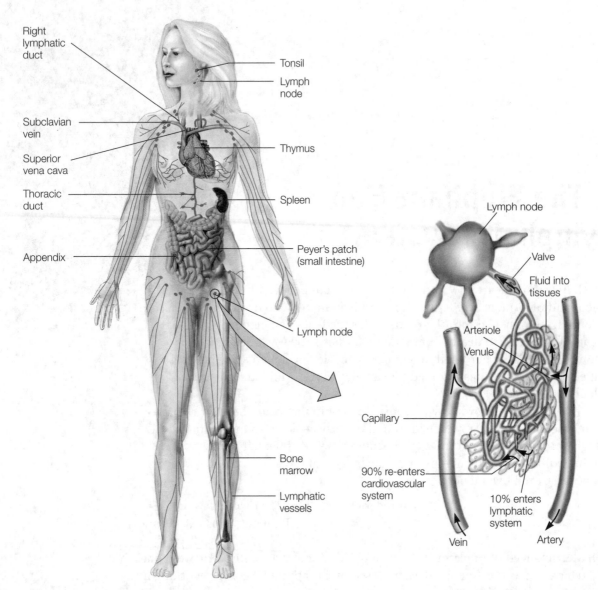

Figure 6.1. The lymphatic system.

2. What are the functions of lymph nodes?

3. If you have a cut on your hand that may be infected, a clinician might palpate your axial lymph nodes (in an armpit). Why would swelling of these lymph nodes indicate an infection?

4. Why would swelling of cervical lymph nodes (in the neck) indicate a sinus infection?

5. Elephantiasis is typically caused by filarial worms blocking the lymph vessels. What would be the immediate consequence for tissue distal to the blockage? If not relieved, how would this result in the deformity described earlier?

INTEGRATIVE CHALLENGE QUESTIONS

6. The lymphatic system also encompasses other organs in the body. Provide a list of other organs that are a part of the lymphatic system and explain their function.

7. Aggressive primary cancers are often capable of metastasizing to other locations throughout the body. One mode by which cancers often spread is through the lymphatic system. Explain the implications for the patient if the tumor did metastasize to the lymph nodes. Also explain why the lymphatic system can provide a means for the cancer to spread to other locations in the body. To which organ systems can cancer metastasize?

(Read more about this system in Daniel Chiras, *Human Biology, 7th edition,* Chapter 3.)

7. Danger Strikes the French Open (Skin and the Integument)

PARIS: July 14, 2009

Kato Fong had made it to the semifinals at Roland Garros Stadium once again. But today he faced his doubles partner, Jacques Clouseau, who had knocked him out of the Australian Open earlier in the year. Fong opened aggressively, taking a 3–0 lead in the first set, but Clouseau came back and almost took the set before tripping on an untied shoelace. Fong ended the set at 7–6.

Fong maintained his momentum and took the first three games of the next set. He was up 4–2 when a fan ran out onto the court waving a gun. As Fong tackled the gunman and wrestled him to the ground, Clouseau pounced on the distraction and aced his opponent, eventually taking the set 7–5.

By the third set, both players were injured and exhausted—but this did not dampen Clouseau's notorious enthusiasm for flirting with fans. As Clouseau blew a kiss to a groupie in a pink leotard, Fong took the set 7–5 and clinched the match. Clouseau pretended to take it all in stride but his attempt to vault the net turned to a fault, and he fell flat on his face.

Kato Fong went on to take the Grand Slam title, destroying top-ranked Charles Dreyfus in straight sets.

1. Both Fong and Clouseau were perspiring throughout the match, despite the mild Paris weather not reaching 70 degrees Fahrenheit (21 degrees Celsius) that day.

   a. Which tissue was generating all the heat in their bodies?

   b. How was that heat transferred to their skin?

c. How does the evaporation of sweat cool the skin?

d. Which molecules are present in their sweat?

e. How do the dissolved particles help draw water out of the sweat glands?

2. The weather in Paris on July 14, 2009, was mild, with humidity of 94%. When Clouseau defeated Fong at the Australian Open earlier that year, Melbourne was in the midst of the worst heat wave on record, with court temperatures approaching 120 degrees Fahrenheit (49 degrees Celsius) and only 10% humidity. The athletes perspired during both matches, but how did the different weather conditions affect their ability to cool down?

3. In an interview with ESPN, Clouseau vehemently denied that women were his downfall once again. He insisted that he was focusing on the ultraviolet danger to those scantily clad fans. But the interview was cut short when he began ranting that the mysteriously protected woman in the pink leotard was part of a global conspiracy.

   a. Which layers of Clouseau's skin are affected by the sun (Figure 7.1)?

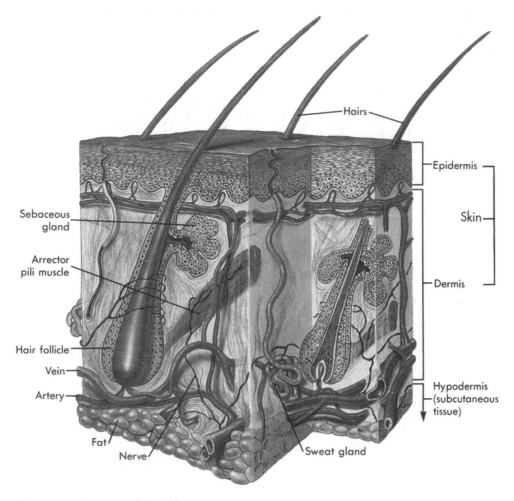

Figure 7.1. Anatomy of the skin.

b. Australians suffer from the highest rates of skin cancer worldwide. How does exposure to the sun lead to skin cancer?

4. Although both athletes were careful to protect their skin, amateur tennis players frequently suffer sunburn early in the season.

a. What is a blister, and why would sunburn lead to a blister?

b. What are the different layers of the integument (Figure 7.1)?

c. How is each of these layers involved in a blister?

5. Between their efforts in tackling the gunman and falling over the net, both players left the court with multiple lacerations.

a. How does skin normally protect the body from infection?

b. Which processes helped the players fight off infection once their skin was torn?

6. In addition to helping cool our bodies, protecting us from the sun, and protecting us from infection, what are some other functions of skin?

7. One of the basic biological principles of physiology is the interrelationship between structure and function. Explain how the architecture of the skin agrees with its function.

8. What role does connective tissue play in skin? What are the different types of connective tissue? Provide some examples of specialized connective tissues in other parts of the body.

(Read more about this system in Daniel Chiras, *Human Biology, 7th edition*, Chapter 4.)

8. Thanksgiving Dinner Distress (Digestive System)

Frank has a large family that always gathers for the holidays. Last year, Frank's family had a full table for Thanksgiving dinner, including aunts, great-aunts, uncles, and even cousins he barely knew. Luckily, Frank's favorite cousin, Colin, a 35-year-old software engineer who works for a toy company in an exotic overseas location, flew in just for Thanksgiving. Colin was happy to be home and was entertaining everyone with his stories of the amazing foods he had eaten over the past few months.

Finally, Thanksgiving dinner was served: turkey, gravy, spinach casserole, sweet potatoes, stuffing, cranberries, lots of good wine, and Aunt Rita's Jello mold for dessert. Just after coffee, Colin doubled over in pain, clutching his abdomen. He said that this reaction had been happening more frequently over the past few months, but he had never felt this bad. His mother wanted to take him to the hospital, but Colin refused, saying he just wanted to lie down on the couch. Meanwhile, Frank's relatives started debating their own diagnoses.

1. Describe the general anatomy of the digestive tract by briefly tracing the path taken by a food particle beginning as it enters the body through the mouth until the waste from the food particle is excreted at the anus (Figure 8.1).

   The food particle enters the mouth and is chewed as the tongue releases salivary amylase, which digests the starch in the food and begins the digestive process. Once the food has been masticated, the food goes down the esophagus, a narrow tube, via peristalsis: muscle contractions. The food then enters the stomach, a sack in the abdominal area in which the digestive enzymes further break down proteins. After the food has been digested into semi-liquid chyme, it passes into the small intestine and mixes with three fluids secreted by the pancreas: bile, pancreatic juice, and intestinal juice. The villi that coat the small intestine absorb the nutrients, which are absorbed into the blood stream. It then travels to the large intestine, which digests complex sugars with villi. When it absorbs enough water, the large intestine sends the waste to the rectum, where it is

2. Aunt Sally said that she was sure that Colin had an ulcer. What is a stomach ulcer? Explain the normal protection of the stomach and describe how this mechanism fails in the case of a stomach ulcer.

   stored until it is released via the contraction of the anus.

   A stomach ulcer is a break in the mucous membrane layer that normally protects the walls of the stomach from the hydrochloric acid that helps digest proteins in the stomach. Ulcers often cause abdominal pain during mealtime. Ulcers are open sores in the stomach lining that bleed into it. H. Pylori, a common bacteria, causes ulcers.

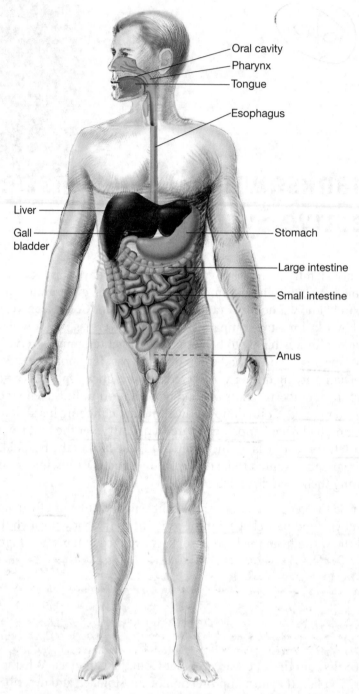

Figure 8.1. The digestive system.

3. What evidence supports Aunt Sally's diagnosis?

- Colin has abdominal pain
- The pain episodes have occurred frequently.
- The pain occurred at the end of a meal when most of the food had been digested, so the acid in the stomach that would be breaking down proteins would be contacting the cherry causing pain.
- Colin works in an exotic place where encountering h pylori might be more likely due to different food relations

CASE STUDIES FOR UNDERSTANDING THE HUMAN BODY

4. Once Colin is feeling better, which questions could Aunt Sally ask Colin to help her figure out if she is correct?

— Has the pain been getting worse?
— Do you get these pains an hour so after you've eaten?
— Have you been eating a large amount of uncooked food, or drinking a lot of alcohol?

5. Cousin Carol said she was sure that Colin was suffering from lactose intolerance. She had the same problem, so she was careful to avoid certain foods.

a. What is lactose intolerance?

Lactose intolerance is a deficiency in the intestinal enzyme lactase that normally can digest lactose. The reduction in the production and secretion of this enzyme does not allow lactose to be digested in the small intestine.

b. Where in the gastrointestinal tract do you usually digest lactose, and why is the pain in a different location from where lactose would be digested?

You digest lactose in the small intestine. The pain is in a different location because the small intestine cannot break down the lactose to absorb the nutrients, so it gets passed on.

6. What evidence supports Carol's diagnosis?

— He is experiencing abdominal pain
— He could have had milk in the coffee

Contra: Eats lots of exotic foods, not typical of someone w/ lactose intolerance.

7. Once Colin is feeling better, which questions could Carol ask Colin to help figure out if she is correct?

— Have you been having irritable bowels/diarrhea? accompanying your abdominal pain?
— Did you have milk with your coffee?
— Do you drink lots of milk in general?

8. Of course, Cousin Linda threw in her two cents by declaring that Colin's problem was a gallstone. Linda knows that gallstones are formed in the gallbladder and are too large to pass through the bile duct. But she is not sure what bile is or why gallstones cause pain only after a person eats certain kinds of meals. Help her explain what gallstones are.

Gallstones are deposits of cholesterol and other materials in the gallbladder that reduce the flow of bile, inhibiting or greatly slowing lipid digestion.

9. What evidence supports Linda's diagnosis?

- There is a lot of fat in the meal they just ate
- The pain has been frequent.
- He works as a software engineer and likes exotic foods, so he could be overweight   - gallstones are common in older overweight adults

10. Once Colin is feeling better, which questions should Linda ask Colin to help her figure out if she is correct?

- Do you eat a lot of fatty foods and feel pain afterwards?
- Does your family have a history of gallstones?
- Are you diabetic?   - Do you have sickle-cell anemia?
     - Do you eat a cholesterol-heavy diet?

11. Rick, Colin's 15-year-old brother, was sure Colin picked up some weird parasite from eating, "all that raw fish and stuff." Explain how gastrointestinal parasites are likely to be acquired. What are some symptoms associated with these parasites?

- Gastrointestinal parasites can enter the body by eating uncooked food, drinking infected water or skin absorption.
     Symptoms include abdominal pain, diarrhea, digestive disturbance

12. What evidence supports Rick's diagnosis?

- Colin experiences abdominal pain
- Colin experiences weakness
- Colin eats a lot of exotic food, some of which may have been uncooked

13. Once Colin is feeling better, which questions should Rick ask Colin to help him figure out if he is correct?

- Do you eat a lot of uncooked food? (exotic)
- ~~How soon aft.~~

14. Colin has been listening to the family discuss his health for the past hour. He sits up and tells the family that he is careful to avoid fat and milk in his diet and the last time he felt this bad was after eating a great meal of spicy Szechuan beef and Hunan duck in hot bean paste. Because nothing in the Thanksgiving meal was fried, he is surprised he had the pain again. Who is most likely to have correctly diagnosed Colin's problem? Explain your reasoning.

not lactose intolerance or gallstones

- Stomach ulcers are caused by H. Pylori bacteria invading the stomach lining, not spicy food.
- These pains Colin experiences occur after meals, which would mean he has raw digesting, which points to a stomach ulcer
- While eating a lot of exotic food might expose you to H. pylori more than "American" food because of differing food production requirements.   - Even though Colin experiences abdominal pain

Sally (ulcer)
~~Carol~~ (lactose)
~~Linda~~ (gallstone)
~~Rick~~ (parasite)

(Read more about this system in Daniel Chiras, *Human Biology*, 7th edition, Chapter 5.)

The food in the second case did not seem to be undercooked, so Sally was right.

In both cases since he felt pain

9. Birthday Party Upset (Liver and Gallbladder)

Adrianne is a Native American living in southern Arizona. At 40 years old, she is expecting her fourth child. Her doctor is somewhat concerned because Adrianne was approximately 45 pounds above her optimal weight before the pregnancy and has already gained 15 more pounds during the first two months of this pregnancy. Recent laboratory tests showed rising blood sugar and cholesterol levels.

Adrianne is trying to watch her diet, but since becoming pregnant she has been having problems with food cravings. One Saturday, her family had a birthday party for her oldest daughter. Adrianne ate several pieces of her favorite pizza—double cheese and pepperoni. That night she was awakened by a sharp pain on the upper-right side of her abdomen. The pain seemed to rise and fall in waves. The pain finally subsided by morning, but Adrianne felt nauseous when she got up. The pain returned suddenly about an hour after lunch. For lunch, she had refried beans, cheese enchiladas, and flour tortillas. Oddly, after the meal, the pain seemed to be behind her right shoulder blade. It soon radiated to her anterior chest to about the level of her collar bone. The nausea returned, and this time was accompanied by two bouts of vomiting.

Adrianne thought that she might have a muscle cramp from trying to lift her youngest son into his car seat the night before, although she had experienced no discomfort at the time. Let's help Adrianne understand what might actually be happening.

1. What are the major organs associated with the right upper quadrant (RUQ) of the abdominal area?

   a. Which technology can be used to easily visualize the anatomy in the RUQ?

2. So far the pain symptoms seem to have arisen soon after a meal and/or possibly after a moderately demanding use of shoulder muscles. If the pain is actually from one cause, what might explain such different pain locations?

3. Let's take a look at the foods Adrianne ate prior to experiencing symptoms. All of them appear to be high in fat, but you know that fat is absorbed into lacteals in the villi of the small intestine lower down in the abdomen. However, when fats first reach the duodenum, their presence stimulates the release of which hormone from the intestinal mucosal cells?

4. Are there any organs in the right upper quadrant that are affected by the release of this hormone? If so, what are those organs and the effects?

5. In addition to detoxifying alcohol (discussed in Chapter 5), what other functions does the liver perform?

6. Bile contains several different components. What are they?

7. Which components of bile can precipitate out when they interact with calcium?

8. Taking these factors into account, summarize why Adrianne feels pain after eating a fatty meal.

9. If left untreated, Adrianne's pain will persist and she could begin to show jaundice. What causes jaundice, and how would her condition be related to it?

(Read more about this system in Daniel Chiras, *Human Biology, 7th edition*, Chapter 5.)

10. Jimi Hendrix and the Digestive Tract (Digestive System)

"There must be some kinda way out of here,"
Said the joker to the thief.
"There's too much confusion,
I can't get no relief.
Businessmen they drink my wine.
Ploughmen dig my earth.
No one will level on the line.
Nobody of it is worth.
Hey!"
All along the watchtower.

—Bob Dylan, "All Along the Watchtower," 1967

Jimmy Hendrix recorded his classic version of "All Along the Watchtower" on his 1968 album, *Electric Ladyland*. Two years later, the joker's wine killed him.

On September 17, 1970, Hendrix drank glass after glass of red wine at a hip party in London. He left the party with his girlfriend, Monika Dannemann, and went back to her flat, where he swallowed a handful of her sleeping pills, Vesperax. The next morning he was dead.

Vesperax is a powerful sedative containing a combination of barbiturates and an antihistamine. Barbiturates are powerful CNS depressants that enhance GABA and inhibit glutamate.

1. What are the general actions of the CNS neurotransmitters GABA and glutamate?

a. At high doses, barbiturates also interfere with the use of calcium at axon terminals. How would this inhibit synaptic transmission in general?

2. Hendrix ingested a combination of barbiturates and alcohol. Most nutrients are absorbed in the small intestine, but alcohol can be absorbed through the mucosa of the stomach.

   a. Which other nutrients can be absorbed before reaching the duodenum?

   b. Why would alcohol have an easy time crossing the mucosa of the stomach?

3. When singer Amy Winehouse was brought to a London emergency room in 2007, the contents of her stomach were pumped to stop further absorption of the drugs she had ingested. Hendrix involuntarily emptied his own stomach (Figure 10.1) by vomiting after a night of drinking. (Nausea and vomiting are triggered by acetaldehyde, the product of alcohol metabolism by the liver enzyme alcohol dehydrogenase.)

   a. To vomit, which muscles must contract?

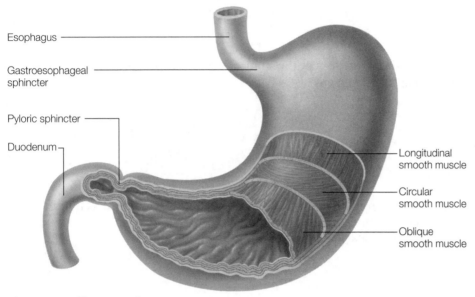

Esophagus

Gastroesophageal sphincter

Pyloric sphincter

Duodenum

Longitudinal smooth muscle

Circular smooth muscle

Oblique smooth muscle

Figure 10.1. The stomach.

   b. Which sphincter normally opens to allow food and drink to enter the stomach?

   c. Which sphincter normally opens to allow the partially digested chyme to move from the stomach into the duodenum?

   d. To vomit, which sphincter must remain closed and which must open during the forceful muscular contraction?

4. Frequent vomiting, or a leaky sphincter, can result in damage to the mucosa of the esophagus.

   a. Which dangerous substances are secreted into your stomach by chief cells and parietal cells to digest your food?

b. How is the mucosa of the stomach better protected from these substances than the esophagus?

c. The mucosa of your duodenum is not as well protected as the stomach mucosa, either. What is mixed into the chyme to neutralize one of its dangerous properties when it enters the duodenum?

5. Although vomiting protected Hendrix from absorbing more alcohol and barbiturates, it ultimately led to his death, when he drowned in the wine that came up. Drowning results from the lungs filling with liquid and blocking the essential exchange of gases. Hendrix did not necessarily inhale wine through his nose; the respiratory tract is connected to both the mouth and the nose.

a. Describe the normal pathway for gases in the respiratory tract and for food and drink in the digestive tract (Figure 10.2).

b. Explain how the glottis and epiglottis normally prevent food or drink from entering the respiratory tract during swallowing.

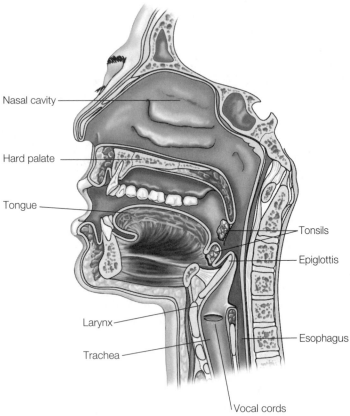

Figure 10.2. The pharynx.

INTEGRATIVE CHALLENGE QUESTION

6. Hendrix's death by drowning in wine was a tragic combination of circumstances. His horizontal position and dulled reflexes contributed to his death. But thousands of *businessmen* suffer a similar fate every year, even when they have not been drinking the *joker's wine*. Why would the businessman be more at risk of choking during a lunch meeting than the ploughman having his lonely meal by the side of the field? Which character is more likely to have ignored his mother's admonition that it is rude to talk with food in your mouth?

(Read more about this system in Daniel Chiras, *Human Biology*, *7th edition*, Chapter 5.)

11. Karen Carpenter (Anorexia)

My body keeps changing my mind

You, you're driving me wild
You're making me fall in love
In love with you baby, I
I'm losing control
I don't wanna let go yet
Till you want me

My body keeps changing my mind

—Karen Carpenter and Richard Carpenter,
"My Body Keeps Changing My Mind," 1979

In her case, Karen Carpenter's mind changed her body so dramatically that it failed. Despite her three Grammy awards, multiple gold and platinum albums, and countless adoring fans, Carpenter felt that her life was out of control. Like many victims of anorexia, she attempted to exert control over her body, which she thought was fat—even when she weighed a mere 80 pounds in 1975. Despite psychotherapy and a healthier diet for some time, Carpenter died 8 years later, at age 32, from the accumulated physiological damage to her body.

1. One major consequence of anorexia is depletion of essential electrolytes, especially calcium and potassium. Almost 90% of women with anorexia experience osteopenia (loss of bone minerals) and 40% have osteoporosis (more advanced loss of bone density).

    a. If Karen Carpenter was not absorbing enough calcium from her diet, which bone cells would digest bone matrix to release the calcium stored there?

b. Which hormone stimulates these cells?

c. Which bone cell type functions in a negative feedback cycle with the afore-mentioned cells?

2. Anemia is commonly associated with anorexia and starvation. Extremely low levels of vitamin B_{12} inhibit the development of blood cells. Direct injection of vitamin B_{12} to the site of hemopoiesis can temporarily increase erythrocyte production. Where are red blood cells formed? Which hormone is responsible for maintaining a homeostatic level of red blood cells in the body? Which protein molecule found in red blood cells is vital for the transport of oxygen throughout the body?

3. Karen Carpenter's fans were shocked by her thinness, but some victims of anorexia and starvation develop bloated abdomens. This is the direct result of the liver's inability to produce blood albumens. Explain how albumens increase the osmolarity of blood and why their absence would lead to edema and bloat-ing. What are some other characteristics of liver dysfunction that victims of anorexia might exhibit?

4. In addition to restricting her caloric intake, Carpenter took laxatives to purge her body of the little food she ate. What role does the colon play in maintaining electrolyte levels and why would laxatives exacerbate any electrolyte imbalance she may already have had? What treatment is normally given to a patient with an electrolyte imbalance?

5. Karen Carpenter also induced vomiting to eliminate the little food she consumed. Bulimics can binge and purge by vomiting many times per day. How would this affect the esophagus?

6. Severe anorexia can also damage nerves and result in seizures and numbness.

   a. How would potassium deficiency inhibit nerve functions?

   b. How would calcium deficiency interfere with synaptic transmission?

7. Karen Carpenter died of heart failure on February 4, 1983. Eight months later, her final album, *Voice of the Heart,* was released. Part of the final problem was that her starving body was digesting muscle, even cardiac muscle, so that it could get protein.

   a. How did her chronically low blood pressure further starve her heart muscle?

   b. How did her low levels of calcium and potassium further inhibit contraction of her heart muscle?

8. Which subcellular organelle is responsible for generating most of the energy needed for the functioning of a cell? Which biochemical (metabolic) processes are actually responsible for generating these energy molecules?

(Read more about this system in Daniel Chiras, *Human Biology, 7th edition*, Chapters 5 and 7.)

12. Aunt May's Heart Attack

Peter couldn't understand why his spider sense was tingling. His wife, Mary Jane, had forgiven him, yet again. His boss, Mr. Jameson, had given him back his job, yet again. And his nemesis, the Goblin, was trapped in a secure vortex, yet again. Peter was shaken from his reverie as a musical chime replaced the tingling on his hip.

"Parker here."

"Peter, where have you been? Aunt May is in the hospital again. She's had another heart attack." Mary Jane's voice had the familiar combination of concern and irritation.

"Sorry, I got stuck at work. I'll swing right over to the hospital."

Peter's Aunt May embodies the essence of unconditional love and has become a beloved icon because of her big heart. But to her cardiologist, a big heart is a dangerous sign of weakening cardiac muscle.

1. Describe how the heart as a muscle does its job of pumping blood. What happens if the cardiac muscle itself does not get enough blood?

2. Using your knowledge of cardiac circulation, explain the flow of blood through the heart (Figure 12.1). What would cause Aunt May's weakness and shortness of breath during her sudden attack?

3. What is the job of the coronary arteries? What happens to their ability to do their job if they become narrowed by fatty or mineralized deposits?

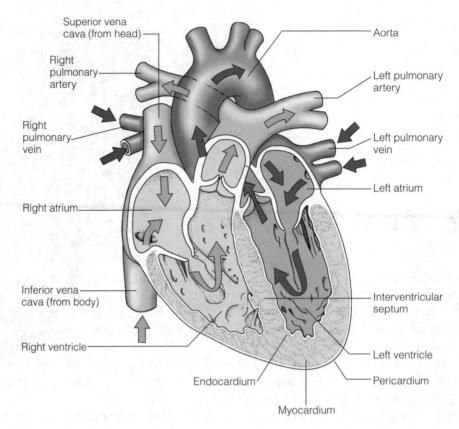

Figure 12.1. Blood flow through the heart.

4. Why are the coronary arteries so easily blocked by floating embolisms?

5. One of the first diagnostic tools used at the hospital was an electrocardiogram (EKG or ECG), which reflects the electrical activity of the cardiac muscle. We know that the atria contract first (the P wave) and then, after a brief delay, the ventricles contract (the QRS complex). Given that the heart does not have any nerves to stimulate the cardiac muscle cells, how is the timing of contraction coordinated? How do action potentials get from muscle cell to muscle cell? If the EKG shows a long delay between the P wave and the QRS complex, which type of cardiac tissue might have been damaged?

6. Before his untimely death, Peter's Uncle Ben was diagnosed with a heart murmur. His primary care physician caught it during a routine checkup. Instead of the normal lub-dub heart sound, she heard a sloshy wobble in place of the lub. The heart sounds are made by the snapping shut of the heart valves. Which of Uncle Ben's valves might have been leaking? Is that sloshy sound heard during atrial or ventricular systole? Is the dub sound heard during atrial or ventricular diastole? The cardiologist jokingly blamed Aunt May for Ben's heart murmur because she "tugged at his heart strings." How did his heart strings—or chordae tendinae—actually support his atrioventricular valves?

7. This time, Aunt May needed quintuple bypass surgery. How does this procedure differ from open-heart valve surgery and from implantation of an artificial pacemaker? Explain the differences between the conditions that these procedures correct.

8. Explain how an angioplasty and the implantation of a stent can also serve as good treatment options for some heart attacks.

INTEGRATIVE CHALLENGE QUESTION

9. Which steps could someone like Peter's boss, J. J. Jameson, take to lower his risk of heart attack?

(Read more about this system in Daniel Chiras, *Human Biology, 7th edition*, Chapter 6.)

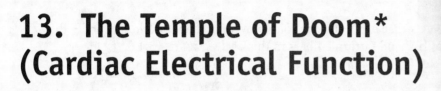

13. The Temple of Doom*
(Cardiac Electrical Function)

Indy and Shorty have discovered the secret passage hidden in the wall behind their guest room in the palace. They grab burning torches and foolishly head off to explore the passageways but before long the two are trapped in a small chamber just as dangerous spikes emerge from the floor and ceiling. They scream, trying to save themselves from the painful death that could await. Just as they are about to be skewered, their sultry blonde companion, Willie, answers their screams and rescues them.

Rather than turning back, the three follow the sound of chanting deeper into the dark abyss of a tunnel. They emerge onto a balcony overlooking an ancient temple hidden beneath their host's palace, gazing in awe and wonder. The mysterious chanting builds in intensity as hypnotized minions call upon the Thuggee goddess of death. Indy, Willie, and Shorty stare in frozen horror as the priest, Mola Ram, plunges his hand into the chest of a slave and removes the beating heart of his conscious victim moments before lowering him into a pit of molten lava, deep beneath the temple floor.

After freeing the slaves and returning the magical stone to its rightful home in the valley, Indy, Willie, and Shorty have time to reflect on what they witnessed.

"That must be some powerful magic, to keep the poor man's heart beating even after Ram pulled it from his chest! Imagine how much it would fetch on the open market," Willie muses. But Shorty knows it didn't require magic. Help him explain how cardiac muscle differs from other muscle in our bodies.

1. What are the three types of muscle? Describe the similarities and the differences among the three types of muscles.

2. Which special feature of cardiac muscle allows an action potential to spread from one cell to another?

---

*Stephen Spielberg, *Indiana Jones and the Temple of Doom*, 1984.

3. Although the autonomic nervous system can affect the rate and intensity of cardiac contraction, Shorty points out that nerves do not directly stimulate cardiac muscle. Which specialized cardiac muscle cells initiate the contraction cycle? Where is this node of cells located?

4. Describe a cardiac cycle and the roles played by the SA node, AV node, the bundle of His, and Purkinje fibers in this cycle (Figure 13.1).

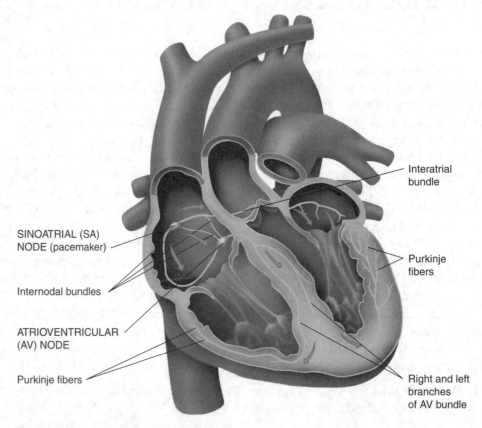

Figure 13.1. Conduction of impulses in the heart.

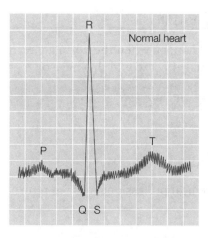

P = atrial depolarization, which triggers atrial contraction.

QRS = depolarization of AV node and conduction of electrical impulse through ventricles. Ventricular contraction begins at R.

T = repolarization of ventricles.

P to R interval = time required for impulses to travel from SA node to ventricles.

Figure 13.2. The electrocardiogram.

5. If Mola Ram were a more sensitive guy, he might have felt the electrical field generated by the coordinated contraction of the cardiac muscle that he held in his hand. This field is actually so strong that it can be sensed by electrodes on your skin and recorded on an EKG. Explain which aspects of a cardiac cycle are visualized in the P, QRS, and T waves of an EKG (Figure 13.2).

6. Why don't we see a wave for atrial diastole in the EKG?

7. Explain what would happen when the normal electrical control of the heart is lost. What detrimental effect would this have on a patient?

8. What devices do medical professionals use to return the normal electrical control of the heart?

(Read more about this system in Daniel Chiras, *Human Biology, 7th edition*, Chapter 6.)

14. Nephritic Syndrome and Edema (Blood and Capillary Exchange)

Your Aunt Rosa has recently been diagnosed with nephritic syndrome, a kidney disorder with symptoms that include high levels of protein in the urine and low albumin concentration in the blood. When you went to visit Aunt Rosa, you encountered her daughter, Maria, who doesn't quite understand the physiology of her mother's illness. Let's help Maria better comprehend her mother's situation.

Maria thinks that her mother's kidney medicine will not help the aching pain her mother feels from the swelling in her legs. Maria remembered that her grandfather got swollen legs whenever he ate salty foods, so she suggested that her mother just needs to stop eating salty foods and perhaps then everything will be fine.

1. Maria is partially correct about her mother's condition being complicated by salt retention, but she needs to know other things about the body's fluids to see the whole picture. The most obvious fluid in the body is the blood. Describe to Maria the contents of blood, including what the plasma contains and all of the different formed elements. Fill in the following table to keep track of what is in the blood:

	Contains/Includes	Functions
Plasma		
Formed elements		
	Erythrocytes	
	Leukocytes	
	Thrombocytes	

2. Maria paid close attention to your description of the components of blood. She turned to Aunt Rosa and declared that instead of just cutting down on salt, Aunt Rosa really needed to cut down on proteins as well, because they are such a major component of blood plasma. Maria reasoned that "Less protein in your diet will lead to less blood and thus to less swelling." At least she is listening to you, but now you should explain to Maria that there are a wide assortment of blood proteins that serve a wide array of functions. Fill in the following table to keep track of what is in the blood plasma. (For now you can skip over the fact that the protein you eat is broken down into amino acids and that blood proteins are assembled in the liver.)

Blood Proteins	Functions
Albumins	
Fibrinogen	
Clotting factors	
Globulins/carriers	
Antigens	
Hormones	

3. Maria admits that cutting out blood proteins could have some severe consequences, but she doesn't see how they affect swelling because she is still working on the assumption that if your aunt reduces her total amount of blood, she can reduce her swelling. Because Maria could be giving her mother very bad advice, she really needs to understand how the blood proteins and other dissolved particles increase the osmolarity of the blood. Then you will be able to explain how blood proteins maintain the osmotic pressure that sucks water into the bloodstream.

Explain osmosis, osmotic pressure, and hydrostatic pressure. Perhaps you can use the diagrams of semipermeable membrane bags of solution immersed in different concentrations of dissolved particles (Figure 14.1) to illustrate these concepts. Use the terms *isotonic*, *hypotonic*, and *hypertonic* as part of your explanation.

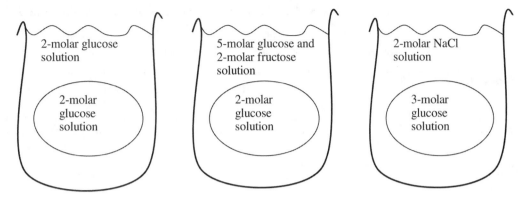

Figure 14.1. Osmosis.

4. Maria listened to your explanation of osmotic and hydrostatic pressure, but she points out that her mother is most disturbed by the general edema caused by the accumulation of fluid in tissues throughout her body. She rephrases her original question: "Isn't it just the high blood volume that leads to increased hydrostatic pressure in the bloodstream and, therefore, to excessive loss of fluid from the blood and its movement into the surrounding tissue?" For her to understand how osmotic and hydrostatic pressure affect the direction of flow into and out of the bloodstream, you will have to explain capillary exchange (Figure 14.2).

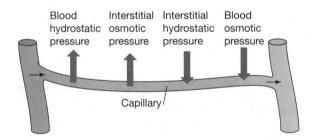

Figure 14.2. Capillary exchange.

The osmotic pressure in the blood is 25 mm Hg, while the surrounding fluid has an osmotic pressure of 3 mm Hg. The blood hydrostatic pressure is 32 mm Hg at the beginning of the capillary bed, but drops to 15 mm Hg at the end of the capillary bed. The surrounding interstitial fluid is under no hydrostatic pressure. The net pressure (NP) reflects the difference in pressure in blood versus in interstitial tissue. The net filtration pressure (NFP) reflects the difference in net hydrostatic pressure (NHP) and net osmotic pressure (NOP). If the NFP number is positive, then the direction of fluid flow is out of the capillaries. Complete this table, and explain why fluid leaves the capillary bed at the arterial end and returns at the venous end of a capillary bed.

	Arterial Side of Capillary Bed		Venous Side of Capillary Bed	
	Hydrostatic Pressure	Osmotic Pressure	Hydrostatic Pressure	Osmotic Pressure
Bloodstream	32	25	15	25
Tissue	0	3	0	3
NP	NHP =	NOP =	NHP =	NOP =
NFP = NHP − P	32 − 22 =			15 − 22 =
Flow Direction				

5. Maria finally admits that just reducing blood volume is not the issue, but that managing the osmolarity of the blood is essential. Even after you explain that the kidneys regulate the concentrations of different dissolved particles in the blood, she asks how this process leads to swelling. The doctor mentioned that because her kidneys are excreting blood proteins that are normally retrieved, the osmotic pressure of Aunt Rosa's blood has dropped to 22 mm Hg. That doesn't sound very different from the normal osmotic pressure of 25 mm Hg. Recalculate the difference between osmotic and hydrostatic pressures in Aunt Rosa's capillaries to demonstrate what a *major* difference this change can make. You can use the following table.

	Arterial Side of Capillary Bed		Venous Side of Capillary Bed	
	Hydrostatic Pressure	Osmotic Pressure	Hydrostatic Pressure	Osmotic Pressure
Bloodstream				
Tissue				
NP	NHP =	NOP =	NHP =	NOP =
NFP				
Flow Direction				

6. After you have explained to Maria how capillary exchange was responsible for her mother's discomfort, she asks a few more questions. Why did the doctor recommend that the nurses relieve Aunt Rosa's pain by massaging her feet and calves? In which direction should her legs be massaged, and why? Why would elevating Aunt Rosa's feet also help?

7. The lymphatic system represents another fluid compartment that Maria should know about. What is the relationship between lymph and interstitial fluid? How is the lymphatic system associated with the circulation and retrieval of proteins?

8. How are lymph vessels similar to veins? How does the lymphatic system return fluids and proteins back into the blood system?

(Read more about this system in Daniel Chiras, *Human Biology, 7th edition*, Chapter 6.)

15. Mountaineering and High-Altitude Risks (Respiratory System)

Tess, who lives in Las Vegas, has been regaling us with her summer adventures in Colorado. She boasted that she climbed four "fourteeners" (mountains more than 14,000 feet in altitude) in the last two summers. She spent two weeks at a former mining town at 9,000 feet, acclimating herself to the environment before undertaking the climbs.

1. What is the main difference in the topography between Las Vegas and Tess's training area?

2. What is happening to Tess's blood as she becomes acclimated to the higher altitude? What is happening to the oxygen-carrying capacity of her blood? Is two weeks long enough for proper acclimation? Explain.

On one of her climbs last year, Tess encountered a party of novice climbers. These inexperienced climbers almost met with disaster. As the two groups climbing together reached 12,880 feet, an athletic young man named Scott became dizzy and fell. He was conscious but was having a great deal of trouble breathing. He had just arrived in Colorado from Kansas City that morning and was upset with himself for possibly ruining the whole vacation. Let's help Tess understand what happened.

This chapter is courtesy of *The American Biology Teacher* (published by the National Association of Biology Teachers).

3. The atmospheric pressure in Kansas City is around 750 mm Hg. The atmospheric pressure at 12,880 feet is approximately 60% of the pressure in Kansas City. Oxygen makes up 21% of the gas in the atmosphere.

   a. What is the partial pressure of oxygen in Kansas City?

   b. What is the partial pressure of oxygen at 12,880 feet?

   c. How would this difference affect Scott's ability to absorb oxygen from the air he breathes?

4. Scott fell on the right side of his chest. He complained of severe pain in the area of his ribs. Let's think about the forces involved in inflating the lungs (Figure 15.1).

   a. If a broken rib pointed inward and punctured a lung, how would that affect Scott's breathing?

   b. If the broken rib suffered a compound fracture and it punctured the thoracic wall, resulting in pneumothorax, how would Scott's breathing be affected?

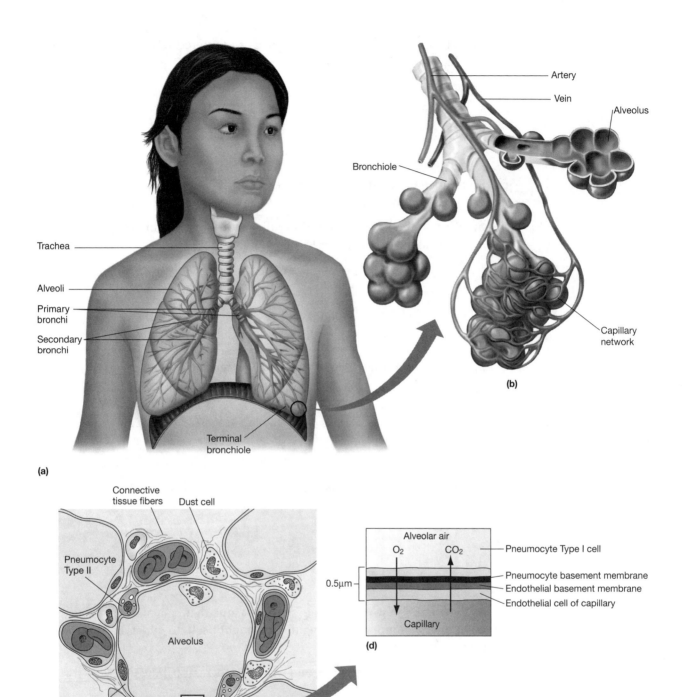

(a)

(b)

(c)

(d)

Figure 15.1. The lungs and bronchi.

c. If he had simply bruised his pectoral and intercostal muscles, how would that affect his breathing?

5. Scott failed to tell his friends that he has exercise-induced asthma. Asthma is characterized by unusual sensitivity to irritation in the airways.

   a. Discuss three ways that asthma affects the conduction passageways of the respiratory tract (Figure 15.1).

   b. Could the altitude have triggered an asthma attack?

   c. How would asthma affect Scott's breathing?

6. As the afternoon wore on, Scott began to panic that he would have to spend the night on the mountain. Fearing that he might begin to hyperventilate, Tess tried to calm him down.

   a. What is hyperventilation, and which factors contribute to it?

b. Explain how the central and peripheral chemoreceptors regulate the rate of respiration.

c. Why is pH an indication of the concentration of CO_2 in the blood?

d. What does the brain monitor to stimulate ventilation?

e. Why might Tess give Scott a paper bag with which to rebreathe CO_2? How could this stop the hyperventilation cycle?

INTEGRATIVE CHALLENGE QUESTIONS

7. When Scott reached the hospital, an x-ray showed that his lungs had begun to fill with fluid. He was given a diuretic to alleviate this problem. Explain why Scott's lungs had begun to fill with fluid. How would a diuretic help?

8. Which muscles are responsible for inhalation? Does exhalation require these muscles as well?

9. What are the alveoli? What function do they serve?

10. What are some other common diseases of the respiratory system? What are some of the treatments used to combat these diseases?

(Read more about this system in Daniel Chiras, *Human Biology, 7th edition*, Chapter 8.)

16. Broken Arrow* (Respiratory Conduction)

Major Vic Deakins, USAF, was not having a good day. His copilot, Riley Hale, kept interfering with Vic's plot to blackmail the government with the nuclear warheads he had just stolen. Riley crashed their $500 million stealth bomber, safely detonated one of the nukes in an abandoned mine, and just kept refusing to die. Deakins still had one nuke, and didn't seem too worried about the helicopter chasing his Hummer through desert canyons. But Pritchett, the guy who bankrolled the whole job, was starting to get on his nerves.

Pritchett: You assured me everything would go smoothly.

Deakins: Everything is going smoothly, I assure you.

Pritchett: It's still my money.

Deakins: And if we succeed, you and your friends will get a ton of it.

Pritchett: *If* we're successful?

Deakins: Look, Mr. Pritchett, I will deliver the weapon to the destination. But I can't guarantee that those jerks in Washington won't do something stupid like . . . not pay.

Pritchett: What if they don't?

Deakins: Well, if they don't, the Southwest will be a quiet neighborhood for, uh, about 10,000 years.

Pritchett: Oh, God! Oh, God! How does that helicopter gunship fit into your well-thought-out strategy? You don't know what you're doing do you? This is out of control! I must have been . . .

Deakins was not having a good day and he was out of patience. He steered the Hummer with his left hand; with his right hand, he grabbed the large metal flashlight from the console between their seats. Before Pritchett could finish his whining, Vic extended his arm and whacked him in the throat. Pritchett grasped his neck, gasped for air, and finally shut up.

Deakins: Hush. Hush!

---

*John Woo, *Broken Arrow,* 1996, 20th Century Fox.

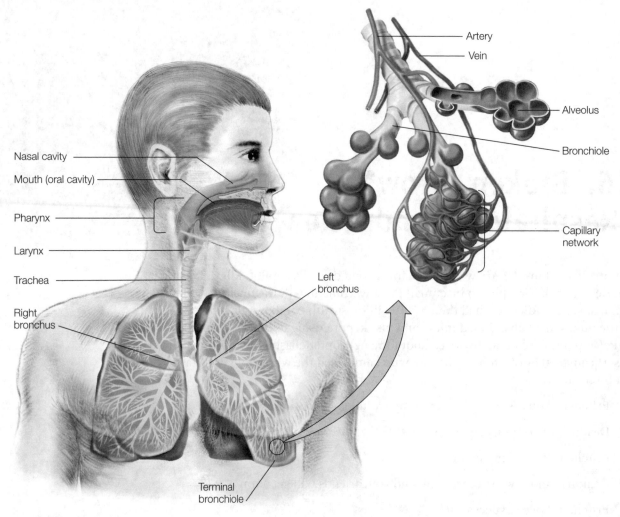

Figure 16.1. Respiratory tree.

1. Despicable as his character was, it was hard not to applaud as Deakins (played by John Travolta) shut Pritchett up. Specifically, which part of Pritchett's respiratory tract did he crush (Figure 16.1)?

2. It seems odd that Pritchett was still able to exhale, but he couldn't inhale.

   a. When we exhale, does the air push against the walls of the respiratory conducting division or pull in? Is exhalation an active or a passive process?

b. Which tissue normally prevents the trachea and bronchial trees from collapsing under the negative pressure of inhalation?

c. Why don't bronchioles need this support?

d. Which muscles are responsible for controlling inhalation?

3. The conducting structures are protected from collapsing during inhalation, but the alveoli are at risk of collapsing at the end of exhalation, even in the dry desert. Explain the force that could cause the wet surfaces of the alveoli to stick together, especially at the end of a long vocal outburst, like yelling or crying.

4. Even though he was whining like a baby, Pritchett was an adult and didn't have to worry about his alveoli collapsing.

a. Which lipoprotein prevents our alveoli from collapsing?

b. Which alveolar cells produce it?

c. Which newborn babies might be at risk of alveolar collapse because they do not yet produce this lipoprotein?

5. Pritchett's crushed airway may have been a dramatic device in the film, but restricted airways are a very real concern for people who suffer from asthma. This threat is so serious that Deakins and Hale would have been disqualified from Air Force flight training if they had asthma.

   a. One trigger for an asthma attack is irritation of the respiratory tract. Which sorts of airborne particles might cause irritation?

   b. Draw a cross section of a bronchiole and describe how inflammation of the irritated airway would restrict airflow.

   c. Asthma attacks can also involve muscle spasm. Redraw your cross section of the bronchiole and explain how smooth muscle spasm would restrict airflow.

   d. Which medications are commonly used to treat asthma? How do these medications aid the person to breathe easier?

6. Explain which measurements can be taken to assess the lung capacity of an individual. Which conclusions can be drawn from knowing the inspiratory reserve volume and the expiratory reserve volume?

7. What function do alveolar macrophages serve? What would happen to the respiration of an individual who has a decreased number of alveolar macrophages due to an immune disorder?

(Read more about this system in Daniel Chiras, *Human Biology, 7th edition*, Chapter 8.)

17. Mrs. Robinson, the Seduction of Smoking, and Cancer

Mrs. Robinson's hair is perfectly coiffed, her gown is shimmering, and an elegant wisp of smoke rises from the tip of the slender cigarette balanced between the tips of her fingers.

"Benjamin, sit down."

"Mrs. Robinson if you don't mind my saying so, this conversation is getting a bit strange."

She smiles calmly and pulls on the cigarette briefly.

"I'm sure Mr. Robinson will be here any minute now . . ."

"No, my husband will be back quite late. He should be gone for several hours."

"Oh my God." Benjamin's anxiety and nervousness become clear now.

"Pardon?"

"Oh no, Mrs. Robinson, oh no."

"What's wrong?"

"Mrs. Robinson you didn't . . . I mean you didn't expect . . ."

"What?"

"I mean you didn't really think I'd do something like that?" Benjamin's awkward nerves escalate as he tries to express himself.

"Like what?"

"What do you think?"   ⟶ has affair w/ girlfriend's mother

"Well, I don't know."

Mrs. Robinson smiles serenely. The hem of her dress is lifted above her thigh as she places her foot on the bar stool next to her.

"Mrs. Robinson, here we are. You got me into your house. You give me a drink. You put on music. Now you start opening up—your personal life—to me and you tell me your husband won't be home for hours."

"So?" she answers innocently.

"Mrs. Robinson, you're trying to seduce me."

She giggles in reply.

"Aren't you?"

—Anne Bancroft and Dustin Hoffman in
The Graduate, 1967 (directed by Mike Nichols)

Anne Bancroft has been described as one of the sexiest smokers on screen, and *The Graduate* is perhaps the best example of her use of a cigarette as a symbol of her sensuality. Ironically, she died of cancer in 2005.

1. Dustin Hoffman has inhaled secondhand smoke on set in many of his films. What is the difference between mainstream smoke and secondhand smoke?

2. Can exposure to secondhand smoke cause cancer? Give an example.

3. There are more than 100 toxic chemicals in cigarette smoke, with 69 having been identified as carcinogenic. Eleven of these components have been associated with cancers in humans, and others are suspected of causing human cancers. The same chemicals are found in exhaled smoke (sidestream smoke), often in greater concentrations. A clue to these greater concentrations has to do with the exhaled smoke being at a lower temperature. How would this difference affect the concentration of these volatile substances?

4. Even though tobacco smoke contains numerous harmful chemicals, nicotine is the one most often mentioned. Nicotine is a vasoconstrictor, a nerve toxin, and a class I insecticide. How might these characteristics affect pulmonary activities and tissue health? Being a vasoconstrictor, how does nicotine influence the healing process during illnesses or following injuries?

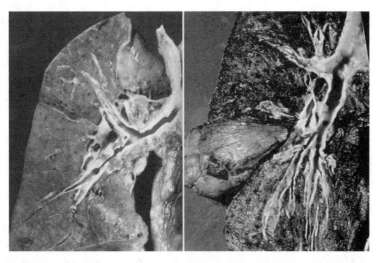

Figure 17.1. The normal and cancerous lung.
Source: Reprinted by the permission of the American Cancer Society, Inc.
All rights reserved.

5. The one word of advice Benjamin got at his graduation party was "plastics." He wouldn't be foolish enough to incinerate plastic and intentionally inhale the fumes, but Benjamin might have been surprised to learn that Mrs. Robinson's cigarette smoke contained cyanide, formaldehyde, arsenic, and ammonium bromide (a toilet cleanser). What are some of the effects of these chemicals on tissues and physiological processes (Figure 17.1)?

6. Lung cancer is not the only cancer associated with smoking. Ninety percent of throat or laryngeal cancers are linked to smoking. What are the daily challenges experienced by patients with end-stage throat cancer? How do some patients have to obtain nourishment? Can they taste their food? What do some of them have to do to speak or communicate?

7. Esophageal and stomach cancers are also associated with smoking. What happens to a person's ability to eat if there is a hole in the esophagus? Which digestive processes are impaired if a person's stomach has to be removed because of cancer? What must patients do to make up for the loss of these processes?

8. Sinus and nasal cancers are also linked to smoking. Smoke damage causes an increase in nasal secretions and tissue swelling. The nasal epithelia lose their cilia and become unable to clean away debris. What happens if the nasal passageways are unable to clean away inhaled debris? Where does this debris go?

9. Most of what we think of as a "flavor" is actually a combination of taste and smell. Chocolate flavor is almost completely due to its smell! We can distinguish five types of tastes but more than 4000 odors. What happens to your sense of smell if tobacco smoke destroys your nasal, olfactory epithelia? How is the ability to maintain good nutrition impaired when you lose your sense of smell?

10. As a student at Berkeley, Elaine might have been interested in the effect of carbon monoxide (CO) in tobacco smoke. Carbon monoxide is another harmful compound in smoke. It is chemically similar to oxygen and binds to hemoglobin more strongly than oxygen. What happens to the body's ability to transport oxygen to cells throughout the body after smoking a cigarette? If Elaine smoked a cigarette just before taking an exam, what would happen to her ability to transport oxygen to her brain when she needed to concentrate her hardest?

(Read more about this system in Daniel Chiras, *Human Biology, 7th edition*, Chapter 8.)

18. Hangover (Renal Regulation and ADH)

Your friend Michelle is not really a drinker, but you two are invited to a wine tasting party on New Year's Eve and decide to go. Michelle preferred the taste of the red wines and sampled many of the offerings—too many, actually. The following morning, she awoke with quite a hangover. Her mouth was dry, and she was very thirsty. She got dizzy when she stood up, and her head was pounding. Michelle closed her bedroom blinds because the light bothered her eyes and made her headache even worse. She remembered having to get up during the night a number of times to urinate.

1. Even though Michelle has been consuming fluids, her mouth is still very dry, and she is quite thirsty. How would you describe her current hydration status?

2. The kidneys play an important role in the body's general hydration by determining how much water will or will not be reabsorbed during urine formation.

   a. Describe the general anatomical organization of the kidney. What are nephrons?

   b. Which part of the brain senses your water balance and stimulates the nephrons to retain or lose water?

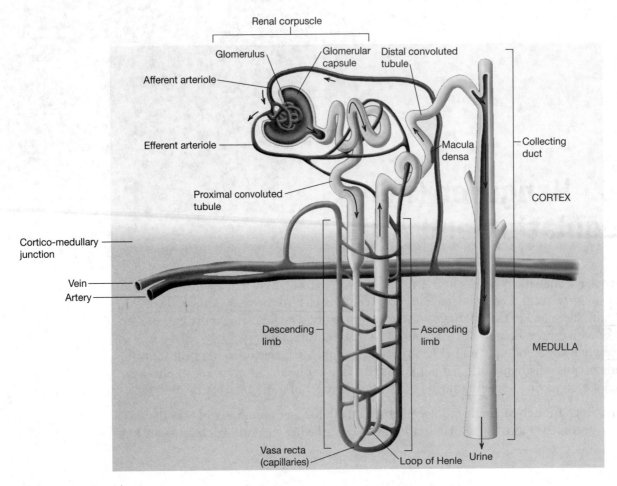

Figure 18.1. A nephron.

3. What is the principal hormone secreted from this part of the brain that will increase the reabsorption of water by the kidney?

4. Michelle's kidneys have increased their urine production even though she has only had a moderate increase in fluid intake. Let's look at what happens in each step of the nephron that could contribute to this symptom (Figure 18.1).

   a. How could increased glomerular filtration lead to increased urine output?

b. Would constriction of the efferent arteriole increase or decrease filtration rate?

c. How does the juxtaglomerular apparatus (JGA) help regulate blood pressure and filtration rate?

5. If the cells of the proximal convoluted tubule (PCT) were inhibited, how would the composition of the urine differ?

a. If the cells of the loop of Henle were not doing their job, would the urine be more or less dilute?

b. If the cells of the distal convoluted tubule (DCT) were inhibited, which urinary contents wouldn't be excreted as efficiently?

c. If the collecting duct became more permeable to water, would Michelle produce more or less urine?

6. The active ingredient in the red wine Michelle was drinking is ethanol. We know that alcohol inhibits the release of the hormone from Question 3, which acts on the collecting duct of the nephron. Can you now explain to Michelle why she feels like she has to keep running to the bathroom?

INTEGRATIVE CHALLENGE QUESTIONS

Let's take a look at Michelle's headache symptoms. Alcoholic beverages—especially red wine—can trigger headaches and even migraines. Alcohol causes intracranial vasodilation and displacement of arteries, veins, and venous sinuses. With Michelle's accompanying water loss, membranes and tissues can also shrink and be displaced.

7. The principal venous sinuses of the brain are located inside the outermost layer of the meninges (the covering of the brain).

   a. What is the name of this layer?

   b. What is the common name given to these venous sinuses?

   c. What do we call the fluid that is enclosed within these sinuses?

8. The outer meningeal layer is anchored anteriorly to the crista galli of the ethmoid bone. With water loss, this membrane shrinks (is displaced) posteriorly. The fluid in its sinuses is compressed. Vasodilation also increases blood flow to many of the meningeal blood vessels.

   a. What happens to the pressure of a fluid that is compressed in a confined space?

b. How might these events relate to Michelle's headache symptoms?

9. If you and Michelle are invited to another wine tasting party, what might you both do to prevent a hangover?

10. Why would coffee be a bad treatment for hangover?

a. What would be a much more useful treatment?

11. What are some of the common diseases of the urinary system?

a. What would cause these diseases?

b. What are some of the treatments used to combat these diseases?

12. If a person were to experience partial renal failure, what are some of the possible treatment options?

    a. If the renal failure becomes more severe, a kidney transplant could become necessary. What are some of the complications of this procedure?

(Read more about this system in Daniel Chiras, *Human Biology, 7th edition*, Chapter 9.)

19. Life in a Dry Environment (Urinary System)

George and his wife, Alice, moved from New Jersey to New Mexico for their retirement. They are both in fairly good health. However, Alice experienced some urinary tract infections from *Escherichia coli*, a bacterium, and sometimes *Candida albicans*, a fungus, right after the onset of menopause. She had fewer infections after her doctor started her on hormone replacement therapy. However, before Alice and George moved to New Mexico, her doctor slowly withdrew her from the hormone therapy because of reported side effects. Alice's doctor recommended that she increase her oral intake of calcium and vitamin D.

1. What hormonal events occur during menopause?

2. What role does the hormone progesterone play in establishing and maintaining the "normal flora" associated with the female reproductive tract?

After several months in New Mexico, both George and Alice noticed that their skin was becoming quite dry, and they were thirsty most of the time. Their urine was also darker in color. Alice began to have some pain in her pelvic and genital areas as well as a dull, burning pain following urination. She decided it was just another bladder infection and tried to drink more water. The pain subsided for a few days but eventually returned with a sharper pain in her abdominal area. She also noticed that her urine was slightly pink in color.

One evening, Alice had a sudden, intense pain in her pelvic area. She had trouble urinating and the fluid was now bright reddish. She went to the emergency room, where healthcare personnel did a urinalysis. The urine sample was cloudy and still tinged with bright red. The results indicated an abnormally high specific gravity, the presence of crystals, microscopic hematuria, and leukocytes; Alice also had a positive test for leukocyte esterase. The pH of her urine was slightly alkaline, and the test for nitrites was positive. The cloudiness, alkalinity, leukocyte esterase, and nitrite findings prompted a bacterial culture, which revealed the presence of *E. coli* and *Proteus vulgaris*. Alice's urine sample also showed elevated levels of magnesium, phosphorous, and calcium, as well as the presence of crystals.

3. You have probably guessed that Alice has a urinary tract infection (UTI). Why are women more likely to have UTIs than men?

4. What are two probable sources of Alice's bacterial infection, and how could they have reached the urinary tract?

5. What does the specific gravity tell you about the urine sample?

6. Because Alice has had sudden, sharp, abdominal and pelvic pain along with the burning sensation following urination, the doctor believes that she may have more than just a UTI. He orders an abdominal X-ray and a noncontrast, helical computed tomography (CT) scan. What do you suppose the doctor suspects?

7. The results of these tests revealed small, highly branching ("staghorn") structures. These structures are consistent with descriptions of structures that are composed of magnesium ammonium phosphate. What are these structures?

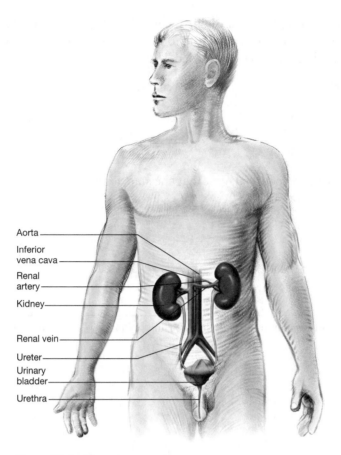

Figure 19.1. The urinary system.

8. These small structures are formed in the kidneys, but Alice thinks she has actually heard them hit the porcelain of the toilet after particularly painful urination. Describe the path they would travel from the kidney to the toilet (Figure 19.1). How does the activity of the ureters add to the pain Alice feels?

INTEGRATIVE CHALLENGE QUESTIONS

9. *E. coli* is the most common cause of urinary tract infections, along with species of *Proteus*, which can split the urea molecule to form nitrites. These organisms can easily colonize the bladder and upper renal areas. They are known to act as "nucleation centers" for the precipitation of excess magnesium and phosphates. Which structures are generally formed around a nucleation center?

10. Before Alice moved to New Mexico, her doctor suggested that she cease taking hormone replacement therapy due to the reported side effects. What are the potential side effects that concerned Alice's doctor?

11. Why did Alice's doctor suggest that she start to increase her oral intake of calcium and vitamin D?

(Read more about this system in Daniel Chiras, *Human Biology, 7th edition*, Chapter 9.)

20. Fast Reflexes (Reflex and Action Potential)

Joe has been on his school's wrestling team since he was a freshman. He does well in the middleweight class, but as he has gotten a bit older, he has put on some weight. Lately his coach has suggested that Joe drop water weight by dosing himself with diuretics just before a meet's weigh-in. While relaxing and watching television one evening, Joe noticed a muscle twitching in his hand. He mentioned this sign to his coach. The twitching was so apparent that the coach suggested Joe go to the campus health services.

When Joe dropped into the clinic the next day, the doctor asked if he had any lifestyle changes. Joe said that he recently upped his running mileage and that he was considering running a marathon. He failed to mention the diuretics!

The doctor listened to Joe's heart and lungs; examined his eyes, ears, and throat; tested the strength of his different muscle groups; and tapped his patellar and Achilles tendons. The doctor asked Joe if he had changed his diet to compensate for the increased activity. Joe admitted that he really was not thinking much about his nutrition, and he still failed to mention anything about the diuretics. The doctor then asked the nurse to draw a blood sample.

1. Why would the doctor tap the patellar tendon? What information could she ascertain from this examination?

2. What would be the normal response to a reflex test?

a. Which type of sensory neuron is stimulated?

b. Using the diagram in Figure 20.1, describe the path of the action potentials along the cells in this reflex arc.

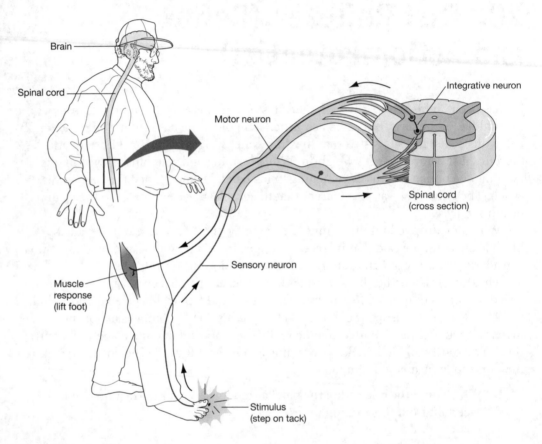

Figure 20.1. A reflex arc.

c. Is it ipsilateral or contralateral?

3. What could a slowed response represent?

    a. What could an increased response represent?

4. What is an action potential? Describe the series of events in the progression of an action potential along these neurons. Which molecules open the ligand-gated sodium channels in the postsynaptic membrane?

5. Describe the movements of sodium and potassium ions during an action potential, and the different voltage-gated channels through which they pass.

    a. What voltage opens the sodium channels?

    b. What voltage opens the potassium channels?

6. If Joe lost a lot of potassium during exercise and before wrestling meets, how would this affect his twitching or reflex speed?

    a. How would Joe's ingestion of diuretics affect his body's levels of potassium?

INTEGRATIVE CHALLENGE QUESTIONS

7. What other questions should his physician ask? Explain how different answers can help determine the diagnosis.

8. Which data might be requested for the lab analysis of the blood sample? How will this information help the doctor understand Joe's symptoms?

9. How does the anatomical organization of nerve cells permit the propagation of nerve impulses? Which special feature of a nerve cell is most important for transmitting impulses with great speed?

10. Many neurodegenerative diseases result in a marked reduction in the patient's reflexes and motor abilities. Identify a few neurodegenerative diseases in which patients exhibit this symptom. Is there a common cause?

(Read more about this system in Daniel Chiras, *Human Biology, 7th edition,* Chapter 10.)

21. Maintaining Blood Pressure (Autonomic Nervous System)

Richard is 58 years old and has been having dizzy spells when he tries to stand up quickly. When he gets out of bed in the morning, he is very unsteady. He can stand for only about 20 minutes before getting dizzy or lightheaded. While standing, Richard's heart seems to be racing. Daily, routine activities are becoming increasingly difficult for him.

One morning Richard tried to get out of bed, and he fainted. His wife called 911. By the time the paramedics arrived, Richard was resting comfortably back in bed. The paramedics decided to take him to the hospital to get checked. When he tried to stand to get on the gurney, Richard fainted again. The paramedics transported him to the hospital, keeping his head and body flat.

Of course, people don't normally faint or feel dizzy when they get up in the morning. The sequence of events enabling you to arise and start your daily routine is somewhat impressive because it takes place in only a few seconds. Let's work through the normal chain of events and see if we can discover in which part of the sequence Richard's problem may lie.

1. The human nervous system is organized into subsystems with different structural and functional characteristics.

   a. How do the central and peripheral nervous systems differ?

   b. How do the somatic and autonomic divisions differ?

c. How do the sympathetic and parasympathetic divisions differ?

2. Circle the correct answers in this flowchart:

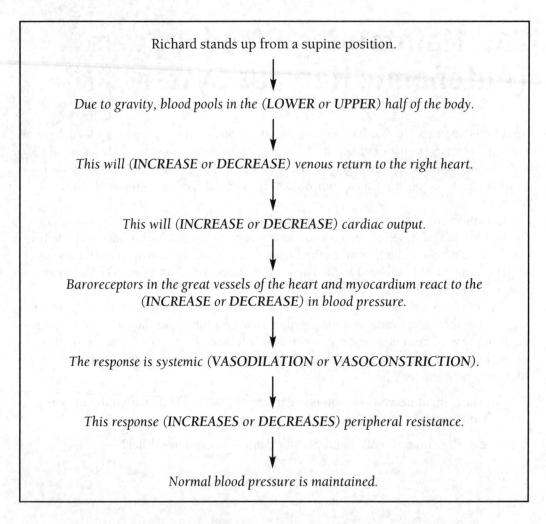

Richard stands up from a supine position.

↓

Due to gravity, blood pools in the (LOWER or UPPER) half of the body.

↓

This will (INCREASE or DECREASE) venous return to the right heart.

↓

This will (INCREASE or DECREASE) cardiac output.

↓

Baroreceptors in the great vessels of the heart and myocardium react to the (INCREASE or DECREASE) in blood pressure.

↓

The response is systemic (VASODILATION or VASOCONSTRICTION).

↓

This response (INCREASES or DECREASES) peripheral resistance.

↓

Normal blood pressure is maintained.

Because Richard experienced fainting when he first stood up, you have probably figured out that his condition has something to do with blood pressure.

3. Where along the previous sequence is Richard most likely to be having a problem? In essence, from what is Richard suffering?

4. Do the neurons of the sympathetic nervous system originate from the cranial nerves and sacral spinal nerves or from the thoracic and lumbar spinal nerves? Where do the parasympathetic neurons originate?

5. The sympathetic nervous system secretes norepinephrine from the postganglionic neurons. What are the effects of this neurotransmitter on the heart and vascular system? Which neurotransmitter is released by the preganglionic sympathetic neurons, the preganglionic and postganglionic parasympathetic neurons, and the somatic motor neurons?

6. Would degeneration of the myelin sheath affect the autonomic nervous system? Are any preganglionic or postganglionic neurons myelinated?

INTEGRATIVE CHALLENGE QUESTIONS

7. While Richard was in the hospital, a major snowstorm occurred. Richard's 65-year-old neighbor, Steve, went outside in the early morning hours to shovel the snow off their shared sidewalk and driveway. Steve meant well, but he is a heavy smoker with high blood cholesterol, and it is hard work shoveling that much snow. Both the cold and the exercise increase sympathetic activity. This will generate body heat, increase heart rate and blood pressure, and deliver more blood to the muscles. But Steve has undiagnosed coronary artery disease and these arteries are now extremely narrowed by the presence of plaque. We have heard stories of people dying from shoveling snow. Discuss what is likely to happen to Steve's heart and why, even though his sympathetic nervous system is responding properly to the initial demands of the situation?

8. Richard's neighbor on the other side, Ronald, is also middle-aged and over-weight. Ronald responded differently to the snow. Rather than shovel snow, he decided to take the day off and stay inside. Watching Steve shovel made Ronald hungry, so Ronald made himself a nice breakfast of eggs, toast, bacon, and waffles. After breakfast, he stretched out in his reading chair with the morning paper and fell asleep. Describe what effect his parasympathetic nervous system is having on Ronald's body during his nap.

9. There are several mechanisms by which blood pressure medications can work. Provide a few examples of medical interventions to lower blood pressure.

(Read more about this system in Daniel Chiras, *Human Biology, 7th edition*, Chapter 10.)

22. High Noon (Reflex)

The long, slow whistle punctuated the deserted streets as the noon train pulled into the station. Frank Miller stepped down, and joined his brother and the rest of their gang. He strapped on his gunbelt, and the four outlaws swaggered down the dusty street to find Marshal Kane and pay him back for sending Frank to prison for five years. Pay him back . . . in lead.

Kane had just married Amy Fowler, the most beautiful woman in the world. She begged him to leave his gun and badge, but his sense of duty wouldn't let him abandon the ungrateful town. He headed towards the train station as the Miller Gang fanned out and surrounded him. The price for Marshal Kane's sense of duty was a bullet in the arm, but not before he gunned down two of Frank's men.

Frank's last henchman, Pierce, was drawing a bead on Kane and was about to shoot the sheriff in the back, when Amy overcame her pacifism and took the gunman out with one shot. The odds were finally in Kane's favor but in the wild west, outlaws don't play by the rules. Frank snuck up behind Amy and took her hostage.

"All right, Kane, come out. Come out or your friend here will get it the way Pierce did."

"I'll come out—let her go," Kane shouted back from the saddle shop across the street.

"Soon as you walk through that door. Come on, I'll hold my fire."

Kane stepped into the doorway but Amy acted quickly and shoved Frank away from her. Before Frank could get his balance, Kane plugged him, putting an end to the Miller gang.

The unworthy townsfolk came out of the woodwork, but Frank tossed his tin star in the dust and the newlyweds finally got to ride off into the sunset.

—*High Noon*, 1952, United Artists (directed by Fred Zinnemann)

1. Gunfighters in the old West were described as having "fast reflexes," but reflex arcs are inherently fast. What are the elements of all reflex arcs?

2. Which part of the central nervous system is *not* involved in a motor reflex?

3. When Kane was shot in the arm, he winced and drew the arm toward his torso. This immediate response to a painful stimulus resulted from a flexor reflex.

    a. Are flexor reflexes ipsilateral or contralateral?

    b. Are they monosynaptic or polysynaptic?

    c. How do flexor reflexes help us avoid injury in everyday life?

4. When Amy pushed Frank away, he stumbled, but didn't fall until he was hit by Kane's bullets. Amy's push didn't knock Frank over because of his crossed extensor reflexes.

    a. Are crossed extensor reflexes ipsilateral or contralateral?

    b. Are they monosynaptic or polysynaptic?

c. How do crossed extensor reflexes help us keep from falling during active sports or even while we get dressed in the morning?

5. Kane, Amy, the Miller gang, and even the cowardly townsfolk depend on stretch reflexes just to walk through the dusty streets of their town.

    a. Are stretch reflexes ipsilateral or contralateral?

    b. Are they monosynaptic or polysynaptic?

    c. How do stretch reflexes contribute to a typical rhythmic walking pace?

6. The stretch reflex of the patellar tendon is a simple diagnostic tool for identifying signaling problems in the peripheral nervous system. What would a slow patellar reflex indicate?

7. Describe the differences between the afferent pathway and the efferent pathway, and explain how each is important for sensation in our bodies.

(Read more about this system in Daniel Chiras, *Human Biology*, *7th edition*, Chapter 10.)

23. Papillon: La Puce's Senses

After a moment he took the bowl and poured the coffee in some mugs. Toussaint leaned down and passed the mugs to the men behind him. La Puce handed me the bowl, saying, "Don't worry. This bowl is only for visitors. No lepers drink from it." I took the bowl and drank, then rested it on my knee. It was then that I noticed a finger stuck to the bowl.

I was just taking this in when La Puce said: "Damn, I've lost another finger. Where the devil is it?"

"It's there," I said, showing him the bowl.

He pulled it off, threw it in the fire and said, "You can go on drinking. I have dry leprosy. I'm disintegrating piece by piece, but I'm not rotting. I'm not contagious."

—Henri Charrière, *Papillon,* 1970, William Morrow and Company, New York

1. Hansen's disease results from a mycobacterial infection of the epidermis, mucous membranes, and nerves. From which developmental tissues are all of these derived?

2. La Puce felt no pain at the loss of his finger. There are several types of sensory mechanoreceptors in the skin that would have already been destroyed by the disease. List the different sensory mechanoreceptors and explain how each functions (Figure 23.1).

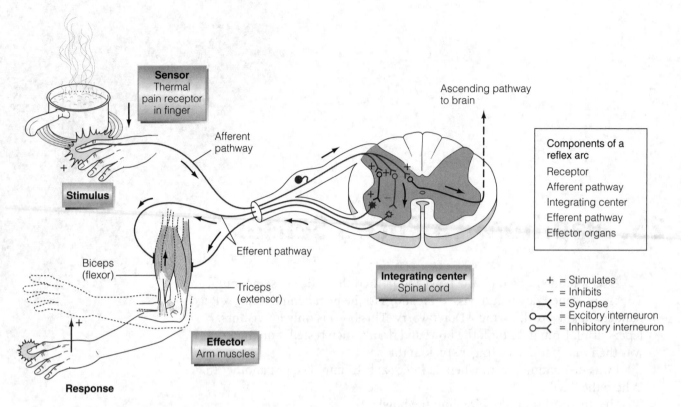

Figure 23.1. Somatic reflex arc.

3. Papillon noticed that lepers had multiple burns on their middle and index fingers where they hold their cigarettes. Several types of sensory receptors are present in the skin, other than mechanoreceptors. List these types of sensory receptors, and explain how each functions.

4. Imagine the reflex arcs in La Puce's body that would be affected by the disease. Describe the differences between the afferent pathway and the efferent pathway, and explain how each is important for sensation in our bodies.

5. As mentioned before, leprosy results from a mycobacterial infection. What are the differences between prokaryotic cells, like bacteria, and the eukaryotic cells of La Puce's body? Also, bacteria can be classified into different categories based on their cellular morphology. Into which classification would you put the bacteria that cause leprosy?

6. Knowing that the presence of foreign bacteria causes this horrible disease, how would you imagine that physicians can treat patients with leprosy? Provide an explanation of which specific mechanism this treatment uses. In other words, which attributes in this treatment ensure that the cells of the patient's body are not grossly impacted, but just the bacterial cells?

(Read more about this system in Daniel Chiras, *Human Biology, 7th edition*, Chapter 10.)

24. Ted Kennedy (Brain)

The work begins anew. The hope rises again. And the dream lives on.

—Senator Edward Kennedy, August 25, 2008

On May 17, 2008, U.S. Senator Ted Kennedy suffered two seizures. Initially, Kennedy's poor cardiovascular health pointed to a stroke, but within days it was announced that the Senator had malignant glioblastoma. On June 2, 2008, surgeons removed much of the tumor from the left parietal lobe of his cerebrum before beginning aggressive radiation and chemotherapy.

Kennedy surprised many colleagues when he arrived on the Senate floor on July 9, to break a filibuster holding up Medicare legislation. However, this was completely in character for the man who had spent much of his 46-year career in the U.S. Senate fighting for the support of cancer research, mental health insurance coverage, and healthcare reform, especially for children. The following month Kennedy energized the Democratic National Convention with his "the dream lives on" speech and repeatedly returned to the Senate for key votes in early 2009.

Over the summer, however, his condition deteriorated. On August 25, 2009, Senator Edward Kennedy died, just months before healthcare reform was passed by Congress.

1. Senator Kennedy had glioblastoma. What are the different types of glial cells, and what are their functions?

2. Which functions would most likely be impaired by damage to the parietal lobe of the cerebrum (Figure 24.1)?

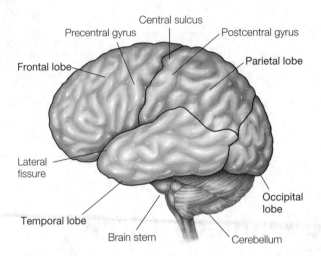

Figure 24.1. Lobes of the brain.

3. Which functions would most likely be impaired by damage to a temporal lobe of the cerebrum?

4. Which functions would most likely be impaired by damage to a frontal lobe of the cerebrum?

5. Which functions would most likely be impaired by damage to an occipital lobe of the cerebrum?

6. The tumor in Senator Kennedy's left parietal lobe affected which side of his body?

7. What is a stroke, and why did Kennedy's poor cardiovascular health and recent heart surgery initially suggest that his seizures resulted from a stroke?

INTEGRATIVE CHALLENGE QUESTIONS

8. What are some of the different technologies that can be used to image the different areas of the brain to determine where a tumor resides?

9. What is the standard course of treatment for a patient with glioblastoma?

(Read more about this system in Daniel Chiras, *Human Biology, 7th edition*, Chapter 10.)

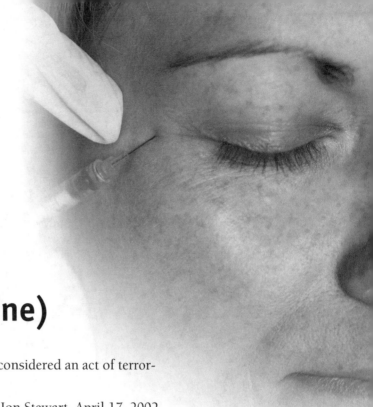

25. Botox (Acetylcholine)

Deadly biological pathogens: When mailed, they're considered an act of terrorism; when injected, the key to eternal youth.

—Jon Stewart, April 17, 2002

The U.S. Senate is considering a bill that would tax Botox. When Botox users heard this, they were horrified. Well, I think they were horrified. It's difficult to tell.

—Craig Ferguson, July 28, 2009

John Kerry was in Florida this week, reaching out and talking with elderly voters. You know, I think it made Kerry a little uncomfortable to be with these elderly people. He finally got a chance to see what he'd look like without Botox.

—Jay Leno, January 16, 2006

The other day at a political fundraiser, Speaker of the House Nancy Pelosi got to meet actor Robert Redford. And witnesses say she was flirting with him. There was an awkward moment when Pelosi winked at Redford and $4000 worth of Botox squirted out.

—Conan O'Brien, October 8, 2009

The botulism toxin (Botox) is the latest weapon in the cosmetic war on natural signs of aging. Washington politicians have joined Hollywood celebrities and millions of Americans who have the muscle-paralyzing toxin injected into their facial wrinkles. Although individual outcomes can vary, the mechanism of the toxin is the same: Botox blocks the release of acetylcholine by motor neurons (Figure 25.1).

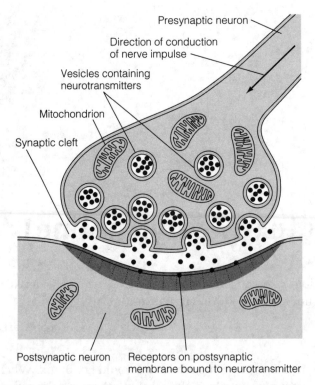

Figure 25.1. Terminal bouton and synaptic transmission.

1. What important functions in your body depend on the neurotransmitter acetyl-choline?

2. Where do neurons store acetylcholine?

   a. What is it that arrives at the axon terminal and causes the release of acetyl-choline?

b. What effect does the release of acetylcholine have on sarcolemma (plasma membrane of muscle cells)?

3. The botulism toxin (Botox) is an enzyme that acts by cleaving the intracellular membrane-bound proteins (SNAP-25 proteins) that are necessary for fusion of synaptic vesicles with the membrane of the axon terminal.

a. Which type of enzyme would you call the botulism toxin?

b. Describe the steps in the release of a neurotransmitter when an action potential reaches the axon terminal under normal conditions.

4. Botulism poisoning often involves ingestion of spoiled food in which the bacteria *Clostridium botulinum* has been growing and secreting the toxin. Paralysis begins in the face and limbs, and can become fatal. Which particular skeletal muscles, if they were paralyzed by botulinum, would stop the victim from breathing?

5. Excess acetylcholine can also be fatal. The venom in the bite of a black widow spider (alpha-latrotoxin) opens calcium channels in the axon membrane. Why would this increase the release of neurotransmitters, including acetylcholine?

6. Alpha-latrotoxin increases release of neurotransmitters, but even under normal physiological conditions, neurotransmitters would build up in synapses if they weren't removed by a variety of mechanisms. Some are reabsorbed by the axon and recycled; others are absorbed across the postsynaptic membrane and digested. A different mechanism is involved in stopping continuous stimulation of the postsynaptic membrane by acetylcholine.

  a. What is the enzyme that digests acetylcholine in the neuromuscular junction?

  b. Sarin nerve gas and the insecticide parathion inhibit this enzyme. In large doses, these enzyme inhibitors are lethal. Why, then, are they useful in small doses to treat Alzheimer's disease?

7. Curare is a South American dart poison that causes paralysis. Unlike Botox, it blocks the acetylcholine receptors.

  a. Where would these receptors be located?

  b. Explain why inhibitors of the enzyme acetylcholinesterase are used as an antidote for curare poisoning.

8. Acetylcholine synapses are found in the cerebrum of the central nervous system, in the somatic nervous system (neuromuscular junctions), and in the autonomic nervous system. However, the neurotransmitter acetylcholine can be either stimulatory or inhibitory, depending on which type of receptors lie on the postsynaptic membrane. Nicotinic acetylcholine receptors on skeletal muscle allow sodium to cross the membrane and initiate an action potential over the sarcolemma. In contrast, muscarinic acetylcholine receptors on cardiac muscle allow potassium to cross the membrane.

   a. In which direction will potassium move when muscarinic channels open?

   b. Will this make the membrane potential more negative, or will it move the membrane potential closer to the threshold voltage?

   c. How would parasympathetic release of acetylcholine affect contraction of cardiac muscle?

9. Acetylcholine is synthesized in the body from acetyl-CoA and choline by the enzyme acetyltransferase. What other important functions does acetyl-CoA serve? What is its role in metabolism?

10. As mentioned earlier, botulism poisoning often involves ingestion of spoiled food in which the bacteria *C. botulinum* has been growing. What measures do you think are taken to prevent this bacterium from growing in the canned goods that we buy from the grocery store? If you were to can the wonderful vegetables that you grow in your garden, which steps would you take to protect yourself from *C. botulinum*?

11. One common theme of the concepts in this chapter is the idea that some substances, such as botulism toxin, can be used as therapies at low concentrations, even though they may be lethal at high concentrations. How does one determine which concentrations are safe to use for therapies and which concentrations are deadly? Which parameters do you think researchers use to answer these questions?

(Read more about this system in Daniel Chiras, *Human Biology, 7th edition*, Chapters 10 and 12.)

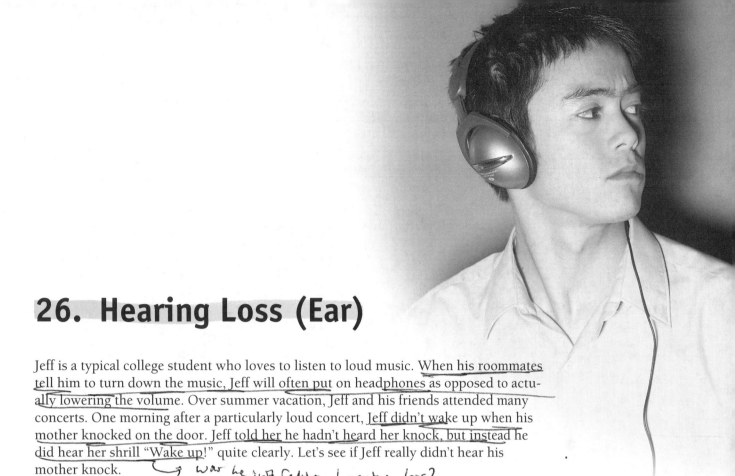

26. Hearing Loss (Ear)

Jeff is a typical college student who loves to listen to loud music. When his roommates tell him to turn down the music, Jeff will often put on headphones as opposed to actually lowering the volume. Over summer vacation, Jeff and his friends attended many concerts. One morning after a particularly loud concert, Jeff didn't wake up when his mother knocked on the door. Jeff told her he hadn't heard her knock, but instead he did hear her shrill "Wake up!" quite clearly. Let's see if Jeff really didn't hear his mother knock. → was he just faking hearing loss?

1. Before we can understand the physiology of the ear, we should focus on the physics of sound waves.

    a. What would the amplitude of the sound wave represent for our hearing?

    b. What would the frequency or the wavelength of the sound wave represent for our hearing?

2. The ear is a series of transducers that convert compressed air waves into nerve impulses that the brain interprets as sounds. What is a transducer?

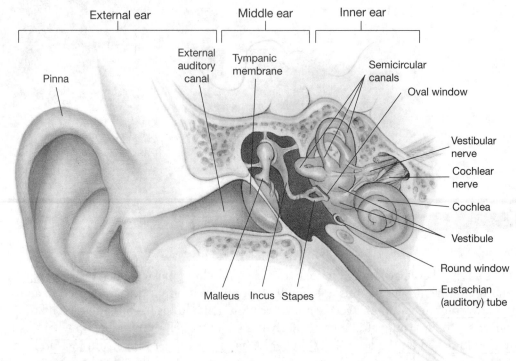

External ear   Middle ear   Inner ear

Pinna

External auditory canal

Tympanic membrane

Semicircular canals

Oval window

Vestibular nerve

Cochlear nerve

Cochlea

Vestibule

Round window

Eustachian (auditory) tube

Malleus   Incus   Stapes

Figure 26.1. The ear.

3. Our ears are organized into three general regions: the outer, middle, and inner ears. Describe each of the specific structures and processes necessary for changing vibrations in the air into messages that can be interpreted by the brain. Start with vibration of the eardrum (tympanic membrane) and work your way through the vibration of the ear ossicles, waves in the fluid in the vestibular and tympanic canals, bumping of the organ of Corti against the tympanic membrane, and ultimately action potentials in the auditory nerve (Figure 26.1).

4. Jeff says that on his CDs, the bass guitar and rhythm tracks seem to be missing.

   a. Where along the cochlear duct does this suggest the damage is specifically located?

   At apex

b. Was he telling his mother the truth about not hearing her knock?

yes, knou will be low freq.

5. Why does it appear that Jeff does not have a torn eardrum? How would he describe his hearing if he had a torn eardrum? Why?

It would be muffled - eardrum is a torn

6. At one particular concert, Jeff and his friends danced and spun one another around. They fell to the floor dizzy and somewhat disoriented.

a. Which part of the inner ear was affected?

Semisircular canal

b. How did the spinning affect this structure and the signals it sent to the brain?

fluid in semi circular canal tells you your monney and eyes

7. Describe what purposes the utricle and the saccule serve for detecting body position and movement.

Semisircular→ acceleration
Other - head position

Utrical is horizontal
saccule is vertical

8. On his flight back to college in the fall, Jeff experienced excruciating pain in his ear. Once the seat belt light was turned off, he called the flight attendant and asked for something to relieve the pain. The attendant brought him a piece of chewing gum instead of the aspirin he expected. To Jeff's surprise, chewing the gum relieved the pain in a few minutes.

   a. What caused the pain in Jeff's ears on the flight?

   b. Was it related to his earlier problems?

   c. Explain how the gum relieved the problem.

   d. Why wouldn't a child with ear tubes suffer this problem when flying in the pressurized cabin of a jetliner?

INTEGRATIVE CHALLENGE QUESTION

9. What are some of the differences between a hearing aid and a cochlear implant?

(Read more about this system in Daniel Chiras, *Human Biology, 7th edition,* Chapter 11.)

27. Detached Retina (Eye)

Rob is the goalie on his soccer team. He is 6 feet 2 inches tall. Rob often must head the ball away from the goal as part of his defensive skills. During a recent tournament, Rob and his team had six difficult games, but they made it to the championship game. Playing the best team from the other bracket, Rob was put to the test. He had to protect the goal time and time again, frequently using his head. After one particularly difficult assault on the goal, Rob signaled to his coach that he wanted out of the game. He said he felt as if a curtain had fallen over his left eye. He could see only half of his normal field.

Rob was immediately taken to the hospital for evaluation. The vision in his right eye was normal, but the vision in his left eye was markedly diminished. He was scheduled for laser surgery the next morning. Rob has no idea how his eye works or what happened.

1. First, let's describe the path of light as it is reflected off the soccer ball until it is an image focused on the retina and then a neural signal on the way to Rob's brain. Describe each structure through which the light passes and what it does to the light (Figure 27.1).

   a. Cornea:

   bends

   b. Iris and pupil:

   iris has muscles that can contract to allow

   c. Lens:

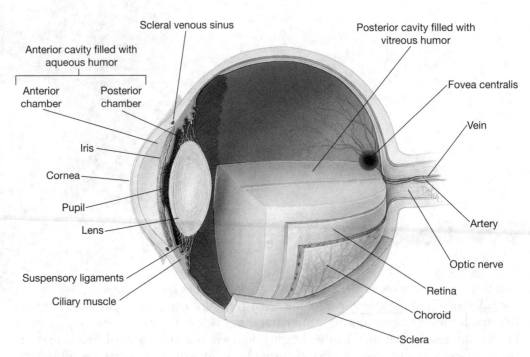

Anterior cavity filled with aqueous humor

Scleral venous sinus

Posterior cavity filled with vitreous humor

Anterior chamber

Posterior chamber

Fovea centralis

Iris

Vein

Cornea

Pupil

Artery

Lens

Optic nerve

Suspensory ligaments

Retina

Ciliary muscle

Choroid

Sclera

Figure 27.1. The eye.

d. Retina:

2. What happened to the retina of Rob's left eye, and why was his vision so quickly and dramatically affected?

Retina was detached from choroid and wasn't receiving nutrents

a. Why is it important to perform the laser surgery as soon as possible? In other words, what does the retina get from the choroid?

b. Why is the choroid so darkly pigmented?

so light doesn't bounce back

3. What are the two types of photoreceptors that exist in the retina?

rods and cones

a. What are the differences between these two cell types?

rods are hypercensitive → better for seeing at nights

cones → color

b. With what nerves do these photoreceptors indirectly synapse?

synapse with bipolar cells then genglior cell

4. Rob wanted to rest, but his family decided to visit him in the hospital. His Aunt Gerty asked why he had to have surgery. Her neighbor had glaucoma that was treated with medication. His grandmother told Rob not to worry about the surgery because she had eye surgery and her vision got much better. She told him she had a new lens implanted. His father said that Rob wouldn't have to wear glasses anymore because the laser surgery would correct his nearsightedness. Compare Rob's condition to those discussed by his relatives and explain how each is different and how each treatment is different.

a. Aunt Gerty's neighbor:

retina would be detached still

b. Rob's grandmother:

His crytaline lens are fine

c. Rob's father:

~~Aut~~ Getting cornea reshaped won't attach retina

d. Rob:

(Read more about this system in Daniel Chiras, *Human Biology, 7th edition,* Chapter 11.)

28. Franklin's Bifocals

By Mr. Dolland's saying that my double spectacles can only serve particular eyes, I doubt he has not been rightly informed of their construction. I imagine it will be found pretty generally true, that the same convexity of glass through which a man sees clearest and best at the distance proper for reading, is not the best for greater distances. I therefore had formerly two pair of spectacles, which I shifted occasionally, as in traveling I sometimes read and often wanted to regard the prospects. Finding this change troublesome and not always sufficiently ready, I had the glasses cut, and half of each kind of associated in the same circle, thus by this means, as I wear my spectacles constantly, I have only to move my eyes up or down as I want to see distinctly far or near, the proper glasses being always ready.

—Benjamin Franklin, letter to George Whatley, May 23, 1785
(Walter Isaacson, Ed., *A Benjamin Franklin Reader,* 2003,
Simon & Schuster, New York)

1. Benjamin Franklin had two different problems with his eyes, requiring two different corrective lenses for different situations. What are the two different structures in our eyes that work as lenses to focus images onto the retina?

2. When he was young, Franklin did not wear corrective lenses. Nonetheless, he was able to change his focus from the words he was reading on a freshly printed broadsheet to the attractive figure of a young lady crossing the street. Which structure in his eyes changed shape so as to change the focus from a close subject to a far subject?

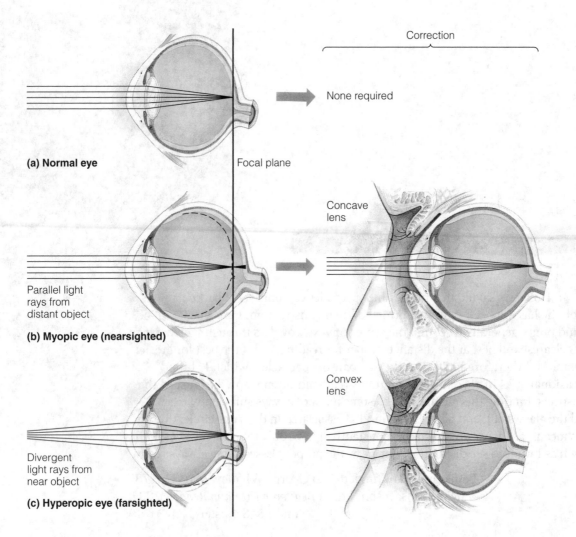

Correction

None required

Focal plane

(a) Normal eye

Concave
lens

Parallel light
rays from
distant object

(b) Myopic eye (nearsighted)

Convex
lens

Divergent
light rays from
near object

(c) Hyperopic eye (farsighted)

Figure 28.1. Common visual problems.

3. To focus on the broadsheet for reading, the crystallin lenses in young Franklin's eyes became thicker and rounder. Explain how contraction of his ciliary muscles relaxes the tension on the ciliary ligaments and allows the lens to bulge into this shape (Figure 28.1).

4. What is an astigmatism? How would this phenomenon affect the manner by which light rays are bent by the cornea?

5. Young Franklin's lens would flatten to focus on the lady across the street. Explain how relaxation of his ciliary muscles changed the tension on the ciliary ligaments to pull the lens into a flatter shape.

6. Even though young Franklin could flatten his lenses, he may still have had difficulty recognizing the woman across the street or focusing clearly on other subjects. How might the shape of his cornea or the length of his eyeball account for this difficulty?

7. Describe the lenses one would wear to correct the focal length of a myopic or a hyperopic eye.

8. Franklin described the fact that he needed reading spectacles, the result of presbyopia. Explain what happened to his aging crystallin lenses that resulted in diminished ability to focus on close objects.

9. Describe the lenses used in reading glasses and how they focus light.

10. Some people have an eyeball that is too short and a retina that is too close to the lens. In that case, do the light rays converge properly in front of the retina or behind it?

11. The curvature of which structure in the eye is altered by contact lenses or LASIK surgery?

12. Because the cornea is avascular, it gets oxygen and eliminates carbon dioxide by diffusion with the surroundings. How are contact lenses made to allow for this gas exchange?

(Read more about this system in Daniel Chiras, *Human Biology, 7th edition,* Chapter 11.)

29. Chocolate (Taste)

And don't call her a cake "decorator," either.

Ursula Argyropoulos is, first and foremost, a pastry chef with impeccable credentials: extensive European training in traditional pastry techniques, experience as a hotel pastry chef, and over ten years as a popular Chef Instructor at Newbury College in Brookline, Massachusetts, which earned her a "Best Educator" award last year from Les Dames d'Escoffier in Boston.

Yet the initial appeal of her work is its extraordinary surface appearance. Ursula's cakes are quite literally works of art: draped in satiny fondant, then graced with a cascade of trompe l'oeil flowers—roses, iris, poppies—whose petals are as finely textured as bone china, whose coloring is as vivid as their botanical counterparts, yet whose fragrance is a barely perceptible whiff of sugar, like the subliminal memory of a sweet dream.

For these are flowers made of gum paste, exquisite handmade recreations of pansies, tulips, lilies, and orchids, caught at the peak of their beauty and adding a breathtaking final touch to Ursula's cakes. The cakes themselves are superb, a richly flavorful marriage of classical European techniques, fine ingredients, and sophisticated combinations of taste and texture.

Not surprisingly, most of Ursula's work is wedding-related. The wedding cake business, she says, is just starting to catch up to the increasing culinary sophistication of the local dining scene.

Reflecting a demographic of older, more worldly brides, these brides-to-be come to her "Art of the Cake" Boston studio with an "awareness that there's more to life than white cake, yellow cake, and chocolate cake," she says.

Although Ursula will "do whatever somebody wants," she will often provide gentle guidance to clients who are overwhelmed by her flavor list: thirteen cake flavors, thirteen fillings, four cake coverings (fondant, marzipan, and white or dark chocolate plastique), twelve buttercreams, and twenty-one "other flavor choices," several available in mousse form.

For those whose eyes glaze over as they browse through options such as pistachio white chocolate cake, black currant mousse, mascarpone Bavarian filling, and walnut buttercream, a page of "recommended combinations" divided under "the nuts" and "the fruits" can have a calming effect.

Of course, it's the flowers that get noticed first. Not only are they astonishingly lifelike—people refuse to believe that the poppies, for instance, aren't "real" until they touch them—but they're made from gum paste, an amalgam that incorporates gelatin and sugar. Most people are more familiar with royal icing, which is made of sugar and egg whites, and is the usual medium for "cake decoration."

Most people, however, are not Ursula. Once she started working with gum paste, it changed her life . . . at least, her professional life.

"I knew I'd be teaching it (gum paste) at Newbury," she says, "so I thought I'd work on it, and as I started, it was like . . . yes! This is it! This is what I want to do for the rest of my life!"

—Excerpt from Fox, Carolyn Faye, 1998, Don't Call Her the Cake Lady, *Gastronoma: Food Files,* Volume I, Issue 1 (Fall 1998)

Ursula's cakes are works of art but, in the end, we enjoy them with our sense of taste.

1. Which class of sensory cells gives us the sense of taste (Figure 29.1)?

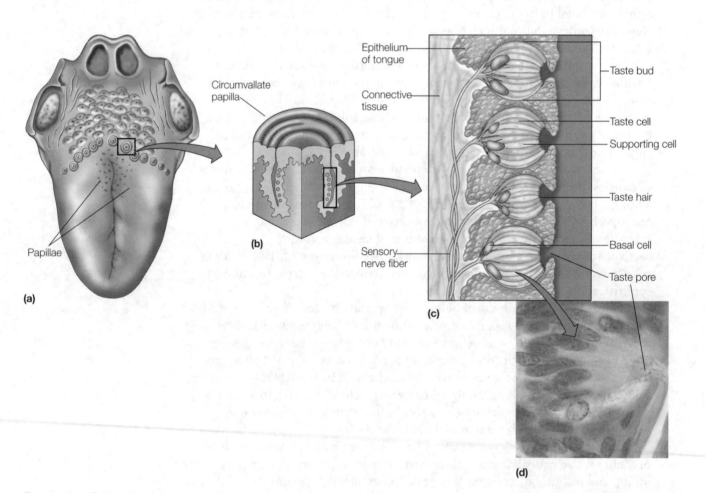

Figure 29.1. Taste receptors.

2. What is the difference between papillae and taste buds?

3. What are the different chemical categories that we can sense in our taste buds?

4. If sensory neurons lit up like Christmas lights when they sent action potentials, which parts of your tongue would light up as you enjoyed one of Ursula's lemon cakes?

5. Which parts would light up as you sipped your coffee or tea between bites of cake?

6. If you burn the tip of your tongue on the soup course, how will this affect your sensation of the taste of Ursula's cake for dessert?

7. If your nasal passage was full of mucus, would you be able to fully enjoy the intense flavor of Ursula's cakes? If not, why? Do we eat with our noses, too?

(Read more about this system in Daniel Chiras, *Human Biology, 7th edition*, Chapter 11.)

30. Chateau Lafite (Sense of Smell)

First tasted in 1985. It was a good cherry red but not as deep as Margaux or Latour; a nose of great depth and immediately forthcoming on the palate. Mouthfilling flavor, a long, hard, tannic finish and notably good aftertaste. Looking over eight notes since then, intensity of color is a feature, the rim changing from its pristine purple to a softer, starting-to-mature red after about five years in bottle. At the same time its bouquet has been gathering momentum. On the last two occasions, the nose has been surprisingly forthcoming, sweet, with strawberry-like fruit, then spice (cinnamon), harmonious, fragrant, lingering. On the palate a soft, rich entry (a feature of all wines made from very ripe grapes, and high in alcohol, a combination associated more with the top California Cabernets). Lovely middle palate, good length, tannin high but well masked by fruit and extract. A fine wine.

> —Michael Broadbent, *The New Great Vintage Wine Book,* 1991,
> Alfred A. Knopf, New York, page 116

Michael Broadbent and other wine masters are able to identify the origin, vintage, and even the particular chateau or winery after sampling a glass of wine. Color, texture, and taste are all clues, but the "nose" of the wine holds the most information. This is why we see wine tasters with their noses deep inside their glasses. But the term "taster" is misleading. Our tastebuds can distinguish only five different chemical properties (sweet, salty, sour, bitter, and meaty). By comparison, our nasal olfactory center can distinguish at least 50 different volatile components of wine and hundreds of other molecules that ought not be found in wine.

So why do Broadbent and other wine tasters slurp and slosh the wine around in their mouths? This action warms the wine and releases heavier volatile molecules into the vapor phase. These compounds reach the olfactory center via the pharynx, where the respiratory and digestive tracts intersect.

1. Where are the olfactory center chemoreceptors located (Figure 30.1)?

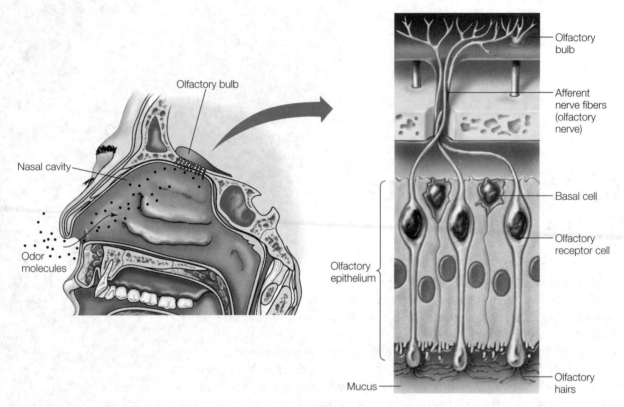

Figure 30.1. Location and structure of the olfactory epithelium.

2. For oxygen to diffuse through the type I alveolar cells and the endothelium to get into your bloodstream, it must first dissolve in the thin layer of water lining the alveoli. What must first happen to an airborne molecule before it can bind to a receptor protein on an olfactory neuron and give Broadbent a hint of strawberry or almond?

3. Why might Broadbent find it more difficult to identify a wine at a tasting in Las Vegas than at a tasting in Seattle?

4. Why would someone with a head cold whose nasal passage is congested complain that he or she is not able to properly taste any food or drink?

5. Why is it that none of the world's greatest wine masters smoke tobacco?

INTEGRATIVE CHALLENGE QUESTIONS

6. Although other sensory neurons synapse to the thalamus, olfactory nerves run directly to the sensory cortex and amygdala.

    a. What does the thalamus do?

    b. What does the amygdala do?

7. Why is it so hard to ignore the smell of the garlic and onion pizza that the guy in the back of the class is eating, but easy to ignore the buzz of the fluorescent lights or the scratchiness of the wool sweater that Aunt Rosa knitted you for Christmas?

8. Why would the nose of their favorite champagne remind Ilse (played by Ingrid Bergman) that she loved Rick (Humphrey Bogart) more than the sound of Sam's (played by Dooley Wilson) piano in the film *Casablanca?*

(Read more about this system in Daniel Chiras, *Human Biology, 7th edition,* Chapter 11.)

31. Myasthenia Gravis (Neuromuscular System)

Shirley was assigned to take Mr. Aletha to physical therapy. She helped Mr. Aletha into the wheelchair and took him down the hallway to the room where he was to receive strengthening exercises for his generalized muscle weakness. Mr. Aletha told Shirley that he often felt okay at the beginning of his therapy session but became weaker as the session progressed. Mr. Aletha was in the hospital because of this perplexing problem.

One day Shirley went to Mr. Aletha's room to take him to therapy, but Mr. Aletha told her that the sessions were canceled: The therapy would not help his condition because he had myasthenia gravis. Shirley had heard about this disease but did not understand the physiology. She went to her anatomy text and began to read the chapters on muscle contraction and nerve transmission. Help her answer the following questions so she can put this puzzling disease in perspective.

1. To understand myasthenia gravis, you have to understand the chain of events from the conscious decision to move a limb to the eventual stimulation of a muscle cell. Describe nerve conduction down the axon of a motor neuron, synapse at the neuromuscular junction, and stimulation of the sarcolemma of one of Mr. Aletha's muscle cells (Figure 31.1).

2. Mr. Aletha has had muscular weakness. Strength depends on individual muscle cells adding their efforts together to generate significant force. Describe the layers of connective tissue surrounding muscle cells, fascicles, and muscle groups that connect muscle cells to one another and ultimately to bones so that their forces can be added together.

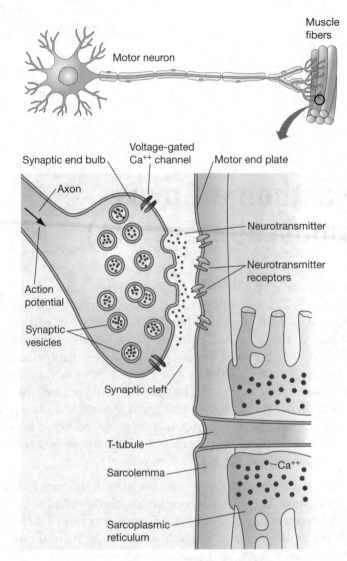

Figure 31.1. A synapse or neuroeffector junction.

3. If you want to lift a whole bag of groceries versus a single apple, how do you increase the force with which you contract a muscle such as your biceps? What is a motor unit? Explain both temporal and spatial summation. What is the neurotransmitter that acts at the motor endplate to initiate muscle contraction?

4. Muscle cells are full of the proteins actin and myosin. Contraction happens when these molecules move past one another.

    a. Briefly describe the arrangement of the actin and myosin molecules within the cell.

    b. How does the interaction between actin and myosin result in shortening of the muscle cell?

    c. Which other molecule provides the energy to break the actin-myosin bond?

5. Myasthenia gravis is an autoimmune disease in which the body's immune cells attack and destroy the acetylcholine (Ach) receptors in the motor end plates of muscles in the shoulder, neck, and face. Why would Mr. Aletha experience a progressive weakening as he tried to do more exercise?

6. How does exercise normally enhance your ability to contract muscles?

    a. What will change in a healthy muscle over a period of months if it is exercised regularly?

b. Why won't exercise help Mr. Aletha?

INTEGRATIVE CHALLENGE QUESTIONS

7. What information does a clinician gain by testing a simple stretch reflex by tapping the patellar tendon below the patella?

    a. Would you expect reflexes of the upper extremities to be any slower in patients with myasthenia gravis?

8. Which types of medications could be used to treat myasthenia gravis patients?

    a. At times, surgical removal of the thymus could potentially be helpful in the treatment of patients with this disease. Why would this be? What role does the thymus have in the immune system, and what consequence would its removal have for an autoimmune disease?

(Read more about this system in Daniel Chiras, *Human Biology, 7th edition,* Chapter 12.)

32. Rigor Mortis (Muscle Contraction)

Emma Louise, RN, started her morning rounds in Mrs. Weitzel's room. Mrs. Weitzel was to have surgery later in the week, and Emma had an order to draw some blood for lab work. Because the patient appeared to be sleeping, Emma took her time tidying up the room. As Emma approached the side of the bed, Mrs. Weitzel suddenly moved her leg. Emma looked over at Mrs. Weitzel and noticed she was not breathing. She punched the code-blue button and checked Mrs. Weitzel's airway to begin artificial respiration and cardiopulmonary resuscitation (CPR).

Several of Emma's colleagues rushed into the room with the crash cart. Mrs. Weitzel was not breathing, was ashen in color, and was very cold. The emergency team could not find a pulse or a heartbeat, and they could not detect respiration. On closer examination, the resident found every joint in her body was stiff.

Just then Mrs. Weitzel's son, Josh, entered the room. He had come to visit his mother and was shocked to see so many people in her room. He couldn't believe his ears when the resident declared his mother deceased. Josh remembered that Emma had been taking care of her and demanded that she explain what had happened.

Emma told Josh that she had entered the room half an hour ago and let Mrs. Weitzel sleep while she tidied up the room. Mrs. Weitzel's son insisted that Emma was negligent because she was standing right there while his mother died. He got so worked up that he even accused Emma of killing his mother. Emma was so upset that she couldn't respond.

Later that week, a board of inquiry convened at the insistence of Mrs. Weitzel's son. Emma was accused of negligence in the care of her patient, but you know this is not the truth. Help Emma explain why Mrs. Weitzel must have been dead before she entered the room.

[handwritten margin notes:] how can ya move a leg if not breathing? → follow that she's been dead for a while → there's a lawsuit to be had

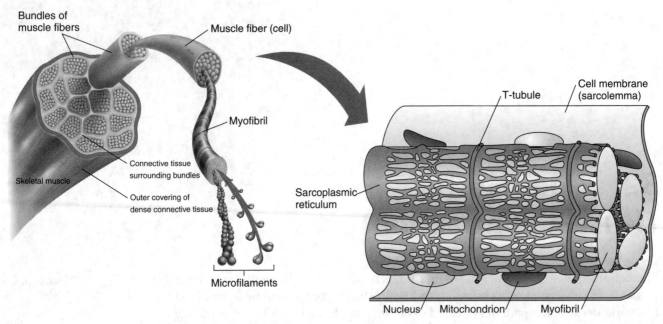

Figure 32.1. Location and fine structure of a muscle fiber.

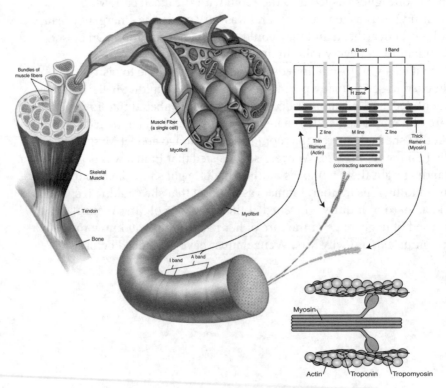

Figure 32.2. Details of the contractile machinery of the muscle cell.

1. Give some background by explaining the chain of events that occur in muscle contraction (Figures 32.1 and 32.2).

   a. Which molecule blocks the myofilaments, actin and myosin, from interacting, and how is it moved when you want to contract a muscle?

   b. What role does calcium play in this regulation, and where is it stored in a muscle fiber?

   c. How is calcium released?

   d. Which external stimulation begins this process?

2. After hearing this explanation, Mrs. Weitzel's son jumped up and complained that tiny molecules of actin and myosin could not be responsible for moving his mother's leg. In response to Josh's claim, explain how actin and myosin are bundled into myofibrils and sarcomeres so that their action is summed to generate significant force in a muscle. What is the overall purpose of the bundling of muscles?

3. Muscle contractions require large quantities of ATP molecules. Which organic molecule is readily hydrolyzed in muscle cells to generate large amounts of ATP?

4. Emma's attorney, Mr. Martin, insists that Mrs. Weitzel was dead before Emma even entered the room.

   a. If Mrs. Weitzel was already dead, how could her leg move? (An action potential did not come down the motor nerves that stimulate the tensor faciae latae.)

   b. Which structures must have become leaky for contraction to occur?

5. Which physiologic process occurred in Mrs. Weitzel's dead body to cause her muscles to lock in the contracted state?

   a. How is it relevant that Mrs. Weitzel was extremely underweight and that Emma had to coax her daily to eat anything at all?

b. How do we usually store energy and oxygen in muscle cells?

6. Take the role of Emma's lawyer, Mr. Martin, at the board of inquiry. Summarize for Mrs. Weitzel's son what must have happened when Emma entered the room. Leave it to the pathologist to explain the real cause of death.

(Read more about this system in Daniel Chiras, *Human Biology, 7th edition,* Chapter 12.)

33. Osteoporosis (Bone Physiology)

Your great Aunt Flora fell and fractured her hip. The doctors said that her hip fracture was between the head of the femur and the greater trochanter—a very common fracture for a 72-year-old woman. You were surprised to hear of this accident because you knew that Aunt Flora was extremely active: She took water aerobic exercise classes three days a week and played golf twice a week.

When you visited your great aunt in the hospital, the doctor told you that she had a severe case of osteoporosis, which probably resulted in the broken hip from the fall. Aunt Flora received a total hip replacement and was up in less than a week. She was anxious to return to her water aerobics class to help strengthen her bones. You knew that she needed other kinds of exercise and treatment as well.

1. Which mineral supplements should Aunt Flora add to her diet?

2. If you were concerned that Aunt Flora had a hormone deficiency, which hormones would you test?

   a. How do these hormones affect osteoblasts and osteoclasts?

3. What exactly do osteoblasts and osteoclasts do in your bones?

    a. How do the hormones you discussed above affect these cells?

4. In addition to water aerobics, which other types of exercises would you encourage Aunt Flora to add to her regimen? Why?

5. Osteoporosis is most commonly thought of as a disease of postmenopausal women.

    a. Why is it also commonly suffered by astronauts after long missions in space?

    b. Why is osteoporosis less common in extremely overweight people here on earth?

INTEGRATIVE CHALLENGE QUESTIONS

6. What would happen to the homeostasis in bone tissue if the feedback relationship between the osteoblasts and the osteoclasts were thrown off, and osteoblasts became hyperactive?

7. How do bone and tendon tissues differ?

    a. In comparison to the healing of a bone, how does the healing of a tendon occur?

(Read more about this system in Daniel Chiras, *Human Biology, 7th edition,* Chapter 12.)

34. Running Knee Injury (Articulation of the Knee Joint)

Ellen is a recreational runner. She averages 20 to 30 miles per week and has been running for 15 years. She participates in occasional marathons but is mostly motivated to run for cardiovascular, pulmonary, and muscular conditioning. In the past few months, Ellen has noticed a persistent clicking sound in her right knee. For some time, she ignored the clicking and continued to run since she had no pain, loss of motion, or strength.

One day, as she was getting ready to run after a warm-up stretch, Ellen felt a pop in her right knee. She could not bear weight on the knee, felt a sharp pain, and was unable to straighten her leg. She crawled into her house. Her husband drove her to the office of their friend David, an orthopedic surgeon. After examining Ellen, David ruled out a torn anterior cruciate ligament by doing the drawer test. He suspected a torn articular cartilage or meniscus. Ellen was sent to a radiologist for an MRI of the knee. This test confirmed David's suspicions and an arthroscopic procedure was performed. Several pieces of torn cartilage, known colloquially as "joint mice," were removed from Ellen's knee. David was pleased to report that all of the ligaments were intact. After a recovery of four weeks, Ellen resumed her training regimen.

1. Describe the knee joint and the movements associated with this type of joint.

2. Which bones make up the knee joint (Figure 34.1)?

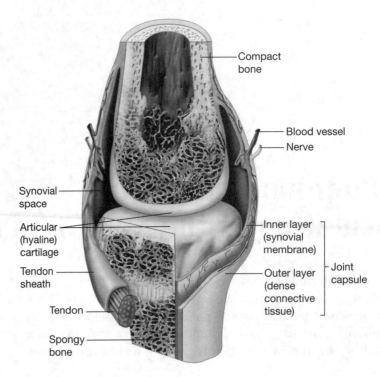

Compact bone

Blood vessel

Nerve

Synovial space

Articular (hyaline) cartilage

Tendon sheath

Tendon

Spongy bone

Inner layer (synovial membrane)

Outer layer (dense connective tissue)

Joint capsule

Figure 34.1. Knee joint.

3. Name and give the function of the cartilages found in the knee.

4. What are the other nonbone structures of this joint?

a. This is a synovial joint. Where is the synovial fluid and what function does it serve?

5. What caused the clicking sounds in Ellen's knee?

    a. Why couldn't she straighten her leg or bear weight on it after the "pop"?

6. What is the function of the cruciate ligaments (anterior and posterior)?

    a. What is the drawer test?

    b. Why do most sports-related injuries involve the anterior cruciate ligament?

7. What would happen if one these ligaments tore?

8. Why is the knee joint the most frequently injured joint?

a. What is another common injury of the knee?

9. What are the menisci in the knee?

a. Where are they located?

b. List some of their functions.

10. Although he doesn't run with her, Ellen's husband understands joint pain. He has dealt with the pain of rheumatoid arthritis for years. How is a joint affected by rheumatoid arthritis?

11. The knee is an example of a movable joint.

a. What are some examples of slightly movable joints in the body?

b. What are some examples of immovable joints in the body?

12. Breaking, or fracturing, a bone is another extremely common injury. Explain the stages involved in repairing a bone fracture.

(Read more about this system in Daniel Chiras, *Human Biology, 7th edition*, Chapter 12.)

35. Jackie Robinson (Diabetes)

On April 15, 1947, Jackie Robinson joined the Brooklyn Dodgers as a first baseman; he was the first African American to play in the major leagues. Although Robinson suffered abusive comments from opposing teams and spectators, friends like Pee Wee Reese and Branch Rickie stood by him. Jewish ballplayer Hank Greenberg partly understood what Robinson was facing. Greenberg told his friend that the best way to deal with racial slurs was just to play better baseball and to win. By the end of his first season, Robinson had earned the Rookie of the Year Award, with 125 runs scored and 29 stolen bases. Robinson's calm demeanor and skill on the field won over fans, but the abuse from opposing clubs seemed to do more to galvanize the Dodgers as a team.

Jackie Robinson played ten seasons in the majors. A year after his retirement, he was diagnosed with diabetes. Despite daily injections of insulin, Robinson lost his → *late onset:* eyesight to diabetes and died of a heart attack at age 53. *your body cells don't respond to insulin*

1. Jackie Robinson's eyes and cardiovascular system were ultimately damaged by uncontrolled blood sugar. ↳ *too much glucagon → to eyes + heart don't think they need blood so starved of it unintentionally, leading to heart attack*

   a. Which hormone is normally released in response to elevated concentration of glucose in the blood?

   b. Which hormone is released in response to low levels of glucose in the blood?

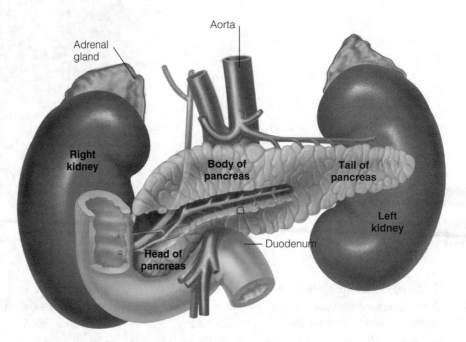

Adrenal gland

Aorta

Right kidney

Body of pancreas

Tail of pancreas

Left kidney

Head of pancreas

Duodenum

Figure 35.1. The pancreas.

2. Which endocrine organ produces these hormones (Figure 35.1)?

a. What are the target cells of these hormones?

3. How do the target cells respond to each of these hormones?

4. Diabetes can result from either an inability to produce insulin (type 1) or insensitivity of target cells to the hormone (type 2). How is each of these problems treated?

INTEGRATIVE CHALLENGE QUESTIONS

5. The brain and other nervous tissue are extremely sensitive to glucose level. These cells can absorb as much glucose as is available, even if there is so much that it damages the tissue. Why do nerve cells have such high energy requirements?

6. The hormones that regulate levels of sugar in the blood are controlled through a feedback loop.

a. Explain the difference between positive and negative feedback loops.

b. Is blood glucose regulated by positive or negative feedback?

7. Which other organ systems are affected by the elevated levels of glucose in a patient with diabetes?

(Read more about this system in Daniel Chiras, *Human Biology, 7th edition*, Chapter 13.)

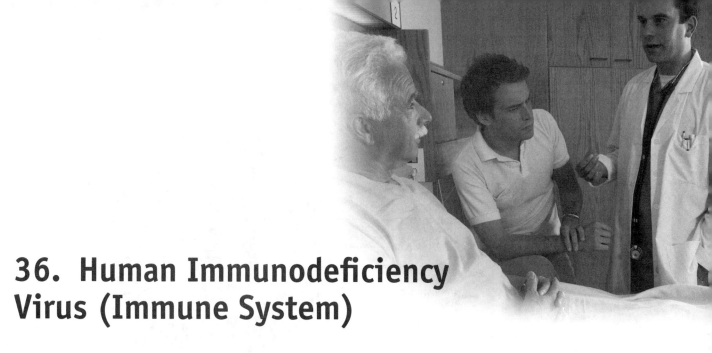

36. Human Immunodeficiency Virus (Immune System)

John's cousin, Albert, had a bad case of pneumonia for several weeks. Albert took antibiotics, but they did not seem to be able to clear up the infection. The doctor finally took a sputum sample, which led to a biopsy of Albert's lung tissue. The tests revealed that the fungus *Aspergillus* caused Albert's pneumonia. Because this fungus is more likely to cause infections in individuals with weakened immune systems, the doctor ordered additional tests. It was soon determined that Albert had acquired immune deficiency syndrome (AIDS).

John was very concerned when he heard this news, because he and his family were around Albert while he was coughing and sneezing from his pneumonia. John is worried that the family might have been exposed to the human immunodeficiency virus (HIV; Figure 36.1). Obviously, John and his family need to know more about how the virus is transmitted and how it harms those who are infected.

Cells have a variety of proteins, glycoproteins, and lipoproteins embedded within their plasma membranes that act as identity markers. These markers serve as targets and/or binding sites for many chemicals, such as hormones and neurotransmitters. They also act as targets and/or binding sites for foreign invaders, including viruses. HIV is able to bind with cells displaying a glycoprotein known as CD4. Examples of such cells include macrophages, monocytes, dendritic cells (Langerhans cells) of the mucosa of the gastrointestinal and genital tracts, uterine cervical cells, neuroglial cells, and helper T lymphocytes.

The monocytes, macrophages, and dendritic cells are part of what is known as the innate or nonspecific immune system. The viruses infecting the dendritic mucosal cells move into local helper T cells. The helper T cells are part of the body's adaptive immune system.

1. What are some of the similarities between the nonspecific and specific immune systems?

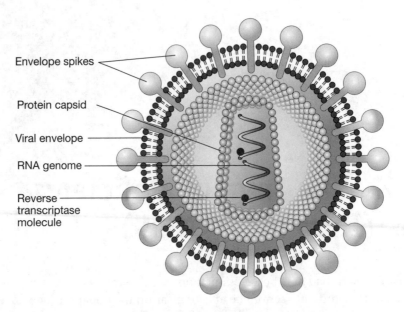

Envelope spikes

Protein capsid

Viral envelope

RNA genome

Reverse
transcriptase
molecule

Figure 36.1. The human immunodeficiency virus.

a. What are the major differences between the two systems?

2. There are many different aspects of the nonspecific immune system. Provide five examples of nonspecific immune defenses.

3. Are any activities of one of these immune systems linked or dependent upon the activities of the other? If the answer is yes, give an example.

HIV is replicated by the cells it has invaded and begins to appear in the blood anywhere from 4 to 11 days after infection. The primary immune response to the presence of the virus comes from cells known as the CD8-positive T cells (lymphocytes). They are also called cytotoxic T cells because they kill virus-infected cells. The cytotoxic T cells are activated by lymphokines secreted by the CD4 helper T cells. They also bind to cells displaying the viral antigenic fragments or epitopes on class I MHC membrane proteins.

4. Which other cells of the innate and specific immune systems do the lymphokines (interleukins) activate?

In many patients, over the course of many years, the virus undergoes many mutations, which continually challenges the cytotoxic T cells to keep up by producing clone after clone of cells. Eventually, too many CD4 helper T cells are killed. The virus also produces a protein known as Nef that represses the synthesis of the class I MHC membrane proteins that present viral fragments to help the immune system identify the virus-containing cells.

5. As the CD4 helper T cells are destroyed, what happens to the overall effectiveness of monocytes and macrophages and other cells of the innate immune system?

6. What effect do declining numbers of helper T cells have on the cytotoxic T cells?

7. What effect does the repression of the class I MHC membrane proteins have on the cytotoxic T cells?

8. How does the loss of the helper T cells affect the activity of B cells (the other major cells of the specific immune system)?

9. Albert had chickenpox as a child and developed active immunity to the *Varicella zoster* virus. Describe the chain of events that gave Albert a population of memory B cells, which could quickly produce plasma cells that could in turn produce antibodies that would recognize the *Varicella zoster* virus.

10. What is the major difference between active immunity and passive immunity?

    a. What are two examples of obtaining passive immunity?

11. Once the immune system starts to "crash," what are some of the principal conditions or diseases to which HIV-infected patients eventually succumb?

    a. HIV genes are not even found in the cancer cells of HIV-associated tumors— so does HIV actually kill people or does it make them susceptible to other diseases that lead to death?

INTEGRATIVE CHALLENGE QUESTIONS

12. What are interferons? Explain how they may protect the body from the spread of a viral infection.

13. Discuss the structure and function relationship of an antibody. Provide some examples of the molecular mechanisms by which antibodies can work.

(Read more about this system in Daniel Chiras, *Human Biology, 7th edition,* Chapter 14.)

37. Sexually Transmitted Diseases (Reproductive System)

Marcia's roommate, Janet, just returned from her appointment at Campus Health Services and was convinced that her life is over. She found out that her ex-fiancé, Brad, had been sleeping around with someone else (hence, Brad's status as "ex"). She is terrified that Brad gave her a sexually transmitted disease (STD) because they had been having unprotected sex. She knows that infections can be treated with antibiotics and wants to start on penicillin right away. To help calm Janet down, first explain to her the differences between bacterial and viral infections.

1. What is the difference between a bacterium and a virus?

2. Which common STDs are bacterial and which are viral?

    a. Would penicillin be equally effective in treating viral STDs and bacterial STDs?

3. Janet is freaking out about every little itch. Let's help her figure out if she has any symptoms of these common STDs. Complete the following table comparing different common STDs.

Disease	Fever	Genital Discharge	Rash	Risk of Infertility	Risk to Neonates	Painful Urination	Mortality Risk
Syphilis							
Gonorrhea							
Chlamydia							
Genital herpes							
Genital warts							
HIV							

4. Some STDs can result in infertility even after the infection is cleared. What happens during healing that interferes with normal fertility?

5. Janet doesn't appear to have any STD-related symptoms, but what can she do in the future to protect herself from contracting or spreading any of these diseases?

(Read more about this system in Daniel Chiras, *Human Biology, 7th edition*, Chapter 21.)

38. Pregnancy Testing (Female Reproductive Cycle)

Farah and Al met while training for triathlons and were just married this summer. Right after their honeymoon bicycle trip through Napa Valley, they returned to Berkeley to begin their respective fellowships in chemistry and physics. Farah was worried because she forgot to take her diaphragm on the honeymoon and she missed her menstrual period the following month. *how is it possible to leave diaphragm at home? Does this refer to something else?*

Farah had already scheduled a gynecological checkup because she has been having irregular cycles for the past few years. The physician drew blood for a complete hormone workup at her first visit. Because of a clerical error, Farah was sent the lab results directly. She brought them into work and has asked you to help her figure out whether she is pregnant.

1. Farah made a list of the hormones on the report. Help her by filling in this table and describing what each hormone does and where it comes from:

Hormone	Gland	Target	Effect	Level
Estrogen				Low
Progesterone			inhibits LH and FSH	Low
Luteinizing hormone (LH)				Low
Follicle-stimulating hormone (FSH)				Low
Gonadotropic-releasing hormone (GnRH)				Low
Human chorionic gonadotropin (HCG)				Undetectable

169

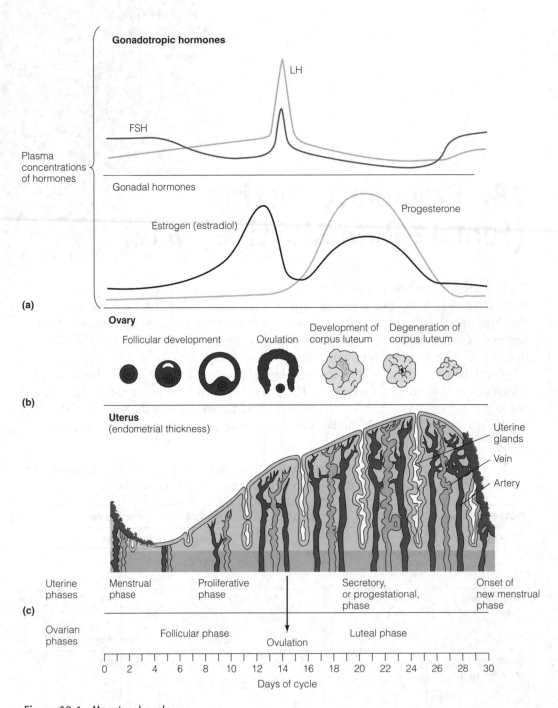

Figure 38.1. Menstrual cycle.

2. To better understand the complexity of these hormones, it is helpful to visualize the changes in these hormones over the course of a menstrual cycle. In a cycle without pregnancy, GnRH, FSH, LH, and estrogen levels all peak just before ovulation. The progesterone level rises and peaks the week after ovulation. Estrogen and progesterone levels then plummet the week prior to menses (Figure 38.1). Help Farah figure out how the different levels of these hormones affect the follicular stages and endometrial changes during the menstrual cycle.

a. Explain the functions of FSH and LH during the follicular phase and the luteal phase of the cycle.

b. Explain the functions of estrogen and progesterone in the menses, proliferative, and secretory phases.

c. How would the levels of estrogen, progesterone, LH, and FSH be different during pregnancy?

3. Farah was going to buy a pregnancy test kit at the pharmacy, but she and Al decided to wait until they received the lab results. For which hormone does a pregnancy test kit test, and why is it a more reliable indication of pregnancy than estrogen or progesterone?

4. Ovulation test kits were on the same shelf at the pharmacy. Which hormone or hormones would indicate ovulation (Figure 38.2)?

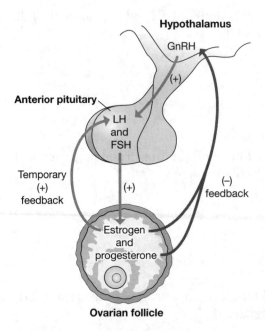

Figure 38.2. Control mechanisms of the hormones of the female reproductive system.

5. If Farah does not want to become pregnant until she and Al finish their fellowships, they might try using the pill or patch for birth control.

   a. Which hormones are present in these contraceptives?

   b. How do they prevent pregnancy?

6. Farah's lab results indicated that she is not pregnant.

   a. Although low levels of FSH and LH, are consistent with pregnancy, which hormone level rules out pregnancy?

b. Which aspects of her lifestyle might explain the low levels of the pituitary hormones FSH and LH?

c. How do they help explain her irregular menstrual cycles?

INTEGRATIVE CHALLENGE QUESTIONS

7. Decreases in levels of estrogen in the body occur when a woman goes through menopause. Explain how the lowered levels of estrogen can affect other systems in the body.

8. Do men have estrogen in their bodies?

a. Do women have testosterone in their bodies?

(Read more about this system in Daniel Chiras, *Human Biology, 7th edition*, Chapter 21.)

39. Infertility (Male Reproductive System)

Lewis and Denisha have recently married and decided to start their family. They have tried for about four months to conceive a child, without success. Denisha called their primary care physician, and he asked that she and Lewis come in together. The doctor explained to them that because the couple are in their early 20s and healthy, the normal procedure is not to worry until they have been enjoying unprotected sex for at least a year. If there is no pregnancy after a year, then they can begin testing for possible causes of infertility.

After a year, Denisha made another appointment because she failed to become pregnant. The doctor met with the couple again, and they discussed some tests. Lewis was surprised to learn that he would also be tested. The doctor told him that in 30% of all infertility cases, the husband has some reproductive-related problem. In an additional 20% of cases, both partners have problems. Thus 50% of all infertility problems actually involve the male partner. Lewis protests that he has "never had a problem performing." Help him understand which other factors are important in determining male fertility.

1. Unlike most diploid cells in the body, sperm cells are haploid cells with only one of each kind of chromosome. Describe the mitotic and meiotic divisions that lead to the production of haploid spermatids from diploid spermatogonia. Be sure to include the intermediate primary spermatocytes and secondary spermatocytes.

2. Spermatids are haploid, but they must still mature into sperm. Describe what happens during spermiogenesis as spermatids complete their development.

3. Why is it important that the germ cells (sperm and eggs) are haploid and not diploid?

4. Ejaculate contains more than just sperm. Describe the contents of semen, and explain which accessory glands produce them (Figure 39.1).

5. The most common cause of infertility in men is a varicocele in the scrotum—that is, a tangle of blood engorged veins (spermatic veins) with improperly functioning or absent valves. A number of theories have been proposed to explain why a varicocele might lead to infertility.

   a. How might venous blood trapped in the spermatic veins produce elevated temperatures in the testicles?

   b. How might slow movement of blood through the testicles result in hypoxia of the germinal epithelium?

INTEGRATIVE CHALLENGE QUESTIONS

6. The motility of the sperm is the most important measure of the quality of the semen. High concentrations of gram-negative bacteria such as *E. coli* can lead to inflammation of the reproductive pathways. Inflammation and other immune responses can cause agglutination of the sperm. What is agglutination, and how might it affect infertility?

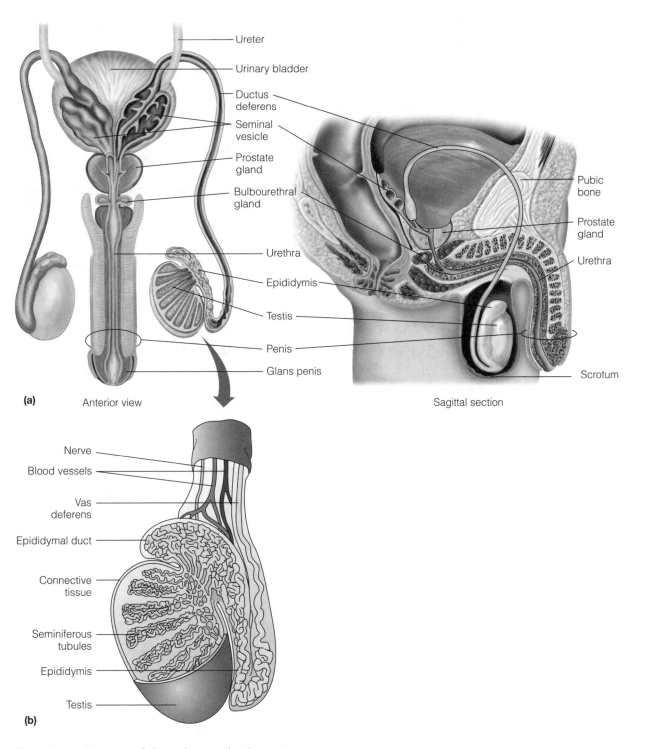

Figure 39.1. Anatomy of the male reproductive system.

7. Chemotherapy and radiation treatments are designed to target highly mitotic cells.

    a. Which reproductive stages are these therapies most likely to affect?

    b. How was the famous cyclist Lance Armstrong able to have children after his extensive cancer treatments?

8. Inguinal hernias are not uncommon in men. What is the inguinal canal, and which events could lead to the formation of an inguinal hernia?

9. The prostate gland is likely to become enlarged as men age. Given the anatomical position of the prostate around the bladder and urethra, which symptoms can occur due to the enlargement of the prostate?

(Read more about this system in Daniel Chiras, *Human Biology, 7th edition,* Chapter 21.)

40. Fragile X Syndrome (Human Genetics)

Mr. Collins just celebrated his 65th birthday. During the past year, he has experienced trembling in his hands and difficulty keeping his balance while walking. His primary care physician began to suspect that Mr. Collins was in the first stages of Parkinson's disease. Over the next several months, Mr. Collins' tremors and poor balance continued, but he did not develop the frozen "mask" facial features so commonly seen with Parkinson's disease.

His physician decided to question Mr. Collins more carefully about his family medical history. The physician thought that obtaining a more complete family history might lead to a better diagnosis, in case Mr. Collins' symptoms were, in fact, due to an inherited disorder. Mr. Collins revealed that his mother was considered to be a "slow learner." He remembered that his father was extremely smart and an avid reader. One of his two sons was diagnosed as severely autistic, while his daughter was diagnosed as having mild learning disabilities. That same daughter has a son who has been diagnosed as learning disabled but not autistic.

The doctor consulted with a neurologist and a behavioral specialist. They decided that genetic studies of the family would be useful because there seems to be a possible inherited link to the conditions seen within the family. The tests showed that the family members have fragile X syndrome. Mr. Collins is concerned that his grandson may pass on this disorder should he decide to have children. Let's help Mr. Collins begin to understand how this condition can be passed from parents to offspring. Fragile X syndrome is a sex-linked, trinucleotide repeat disorder involving extra copies of CGG sequences. It is the most common cause of inherited mental disorders associated with mental retardation and autistic-like behaviors. A large range of signs and symptoms associated with physical, cognitive, intellectual, social, and behavioral abnormalities are described, with some being more distinct in one sex or the other. Signs and symptoms generally manifest from very early childhood, even in infancy. However, older males who carry the gene can suddenly develop symptoms that are characterized as fragile X-associated syndrome, including Parkinson's-like symptoms of loss of muscle coordination, poor balance, and disorganized gaits. Tremors are also common.

The inheritance of this disease is quite complicated. The gene carrying the repeat sequences tends to become unstable over time. A gene carrying 60 to 200 copies of the CGG repeat is known as a premutation; a gene with more than 200 copies is known as a full mutation. In females, the premutations can expand into full mutations in some cells, including the reproductive cells. Males can carry the premutation form and have no symptoms, unless they occur later in life and are manifested as Parkinson's-like symptoms. Men pass this gene on to their offspring as the premutation. Males inheriting the full mutation almost always have fragile X syndrome. Females who have inherited the premutation sometimes show symptoms known as premature ovarian failure (POF), which can lead to infertility and early menopause. Sixty-five to seventy percent of females with the full mutation show some signs of fragile X syndrome, albeit usually not as severe as the signs in males who have the full mutation.

1. Draw a family tree like the one in Figure 40.1 for Mr. Collins. Place his parents, Mr. Collins, his sons, his daughter, and his grandson on the tree. Put appropriate X and Y chromosomes on each family member.

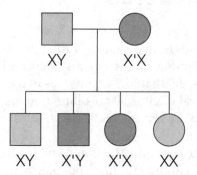

Figure 40.1. Inheritance of a sex-linked dominant gene pedigree.

2. Given the phenotypes of Mr. Collins' children, what can you deduce about the genotype of Mr. Collins' wife?

3. Explain how a gene on the X chromosome might be inherited "differently" among male and female descendants.

4. Males inheriting a full mutation will almost always have the serious symptoms of fragile X syndrome. Explain how this phenomenon is related to this syndrome being a sex-linked disease.

5. Not all females inheriting the full mutation show signs of the disease. Even those who do show signs typically have milder symptoms than males. How is this phenomenon related to fragile X syndrome being a sex-linked disease?

6. Will a male who has the full mutation pass the disease along to his sons? Explain.

7. What is the likelihood that a male with the premutation will pass the disease along to his daughters? Explain.

8. If a male passes the premutation along to his daughter, is she likely to have fragile X syndrome? Explain.

9. Could any of Mr. Collins' grandchildren, through his daughters, have the disease? Explain.

10. Could any of Mr. Collins' grandchildren, through his sons, have the disease? Explain.

(Read more about this system in Daniel Chiras, *Human Biology, 7th edition,* Chapter 17.)

41. Prenatal Care (Development)

Marcia is a 19-year-old who is away from home for the first time. She is attending a large university and is pledging a very prestigious sorority. Marcia drinks to be social with her friends in the evening three to four times a week. She recently met a nice guy at one of the social events, and they began dating on a regular basis. She and her new boyfriend spent their holiday break together. About three months later, Marcia discovered that she was pregnant.

Marcia chose to continue the pregnancy and made an appointment to see a nurse practitioner at the campus health center. This nurse was a bit conservative with her philosophy regarding alcohol consumption during pregnancy. The nurse told Marcia that scientists have not determined that there is any safe amount of alcohol consumption during pregnancy. She also told Marcia that drinking small amounts of alcohol infrequently appears to have less effect on pregnancy than drinking large amounts at a time. Marcia assured her that she has only an occasional glass of wine and does not engage in binge drinking.

The nurse practitioner told Marcia that alcohol has the greatest effects on the fetus during its early development. When she is drinking, the mother's blood alcohol level rises. Alcohol, because it is a small molecule, readily crosses the placenta and enters the fetal circulation by way of the umbilical cord.

The nurse practitioner encouraged Marcia to see an obstetrician and scheduled her for regular follow-up visits at the health center. She also encouraged Marcia to call her if she has any questions or concerns. Let's help Marcia understand the information offered by the nurse practitioner.

1. Describe the major developmental events taking place in the fetus during the following time intervals. Indicate for each time interval what damage could occur because of maternal alcohol consumption. It is during these weeks that a woman may not know she is pregnant and continues to drink.

	Developmental Events	Possible Damage
Weeks 1 and 2		
Weeks 3 and 4		

	Developmental Events	Possible Damage
Weeks 5 to 8		
Weeks 9 to 12		

2. How does caffeine or cocaine affect a developing fetus?

3. Why is it important for mothers to consume sufficient amounts of folic acid in their diets early in pregnancy?

4. Should Marcia schedule an ultrasound when it is appropriate? Why?

INTEGRATIVE CHALLENGE QUESTIONS

5. What are some of the hormones that are produced by the placenta? What functions do each of these hormones serve?

6. Provide some examples of tissues that arise from each of the three embryonic germ layers.

(Read more about this system in Daniel Chiras, *Human Biology, 7th edition,* Chapter 22.)

CASE STUDIES FOR UNDERSTANDING THE HUMAN BODY

42. Drug Resistance (Natural Selection)

Sylvia and her boyfriend, Ricardo, called in "sick" and took the day off to go on a picnic in mid-July. Sylvia packed a nice lunch of potato salad, chicken salad sandwiches, chips, coleslaw, and homemade ice cream. The pair ate around noon and then joined some other couples for a game of touch football. About five o'clock they snacked on the leftovers, which by now were a little warm.

About seven o'clock the next morning, Sylvia awoke with abdominal cramps. She had two bouts of diarrhea and took some nonprescription medication to relieve her symptoms. Ricardo began experiencing the same symptoms about three hours later. Even with the medication, Sylvia's diarrhea continued into a second day, accompanied by abdominal pain, cramping, and a fever. Her mother had some extra cephalosporin antibiotic from a previous illness so she gave it to Sylvia, who took it for five days. Her symptoms had actually subsided the evening of the second day from their onset.

Several months later, Sylvia had a ruptured appendix and developed an infection. After surgery, she was placed on cephalosporin—the antibiotic that is traditionally effective in these cases. The infection did not respond to the antibiotic, and she was placed on another one. Her infection still did not respond, and two more antibiotics were given before one proved successful in clearing up the infection.

Sylvia thought her inability to clear the infection was punishment for skipping work. Ricardo didn't disagree that it might have been partly her fault, but he said she helped the bacteria in her gut evolve resistance to antibiotics. Sylvia thought that idea was ridiculous and that Ricardo was not doing his job of being sympathetic and supportive. Let's help Ricardo explain.

1. Which bacterial food poisoning is most likely when a person eats chicken, mayonnaise, ice cream, or other egg products that have been left unrefrigerated?

a. When Sylvia took the cephalosporin, it killed 99% of the bacteria in the first three days. But then she stopped taking the medication because her mother had given her only a few leftover pills. How did the remaining 1% of the bacteria differ from the 99% that died?

2. Which bacterium is the most likely candidate for causing the peritoneal infection from Sylvia's ruptured appendix?

3. What are transposons?

a. How are transposons produced and passed among bacteria?

4. What are plasmids?

a. How are plasmids produced and passed among bacteria?

5. Which type of genetic information is most likely to be carried by transposons and plasmids?

6. In a bacterial population, are all the individual bacteria genetically identical? Explain.

7. How can a bacterial subpopulation become the predominant "normal flora" population?

8. The infection from the ruptured appendix was probably caused by one of Sylvia's normal flora (a gram-negative rod). It was resistant to several antibiotics. How could Sylvia's "normal" bacterial population acquire drug resistance to the antibiotic she had taken, cephalosporin, as well as other antibiotics that she had never taken?

9. Do you think Sylvia's multiresistant bacterial population arose because of the activity of plasmids, transposons, or both? Explain.

10. How can we help prevent such potentially harmful bacterial populations from becoming the dominant flora in an individual?

(Read more about this system in Daniel Chiras, *Human Biology, 7th edition,* Chapter 23.)

43. Alternative and Herbal Remedies (Scientific Evidence and Decision Making)

Donna is a 20-year-old college student. She lives off campus in an apartment with three other students. Donna has had a cough and sore throat for two weeks. She says it is painful to swallow any liquids, but she is very careful to take her gingko biloba pills three times a day, along with echinacea tea and St. John's wort. Her friend told her that the gingko biloba helps balance the good and bad immune cells. Donna also read an article about a study that showed St. John's wort inhibits the growth of fungi and viruses. She also read that the echinacea helps balance her hormones, reduces her monthly cramps, and fights acne.

Donna's roommate, Eunice, is taking microbiology this term. She just started reading about gram-positive bacteria and suggested that Donna's symptoms sound like strep throat. She wanted to swab Donna's throat and culture it to test for *Streptococcus*. Donna was skeptical of Western science, stating that is "no more scientific than the much older Eastern medical traditions." However, she was willing to help her roommate out because Eunice can get extra credit for identifying "wild" samples.

Two days later, Eunice came home from class and found Donna bravely downing her tea. She told Donna that the culture tested positive for strep and that she should go see the campus nurse and get a prescription for an antibiotic to clear the infection. Donna resisted her friend's suggestion, saying that she did not want to take antibiotics, which she views as "just chemicals" that will "poison her body" and "support some big drug company." She would rather take something "natural." Is there any way to convince Donna that the antibiotics are the best treatment for strep?

1. Because gingko biloba, St. John's wort, and echinacea are not regulated by the FDA, what regulations do the sellers of those products have to follow? How much testing is required before the manufacturers can claim benefits on the package of the product?

2. Donna is more suspicious of the pharmaceutical industry than the herbal remedy industry. Are herbal remedies sold on a not-for-profit basis?

3. What is the difference between Eunice saying that antibiotics will clear Donna's infection and Donna's other friends telling her that St. John's wort will clear it?

4. What is wrong with the following argument: "Because more than half of the prescribed drugs come from botanical products, natural products like gingko biloba are just as good and less processed."

5. Donna pointed out that her friend Sheana had a sore throat and got better after taking twice her normal daily dose of St. John's wort. What is wrong with this argument as evidence of the efficacy of St. John's wort? (Whether St. John's wort works or not, there are a number of serious flaws in Donna's line of reasoning.)

6. Because the infection didn't go away, Donna wanted to try star anise. She read that it can cure rheumatism, colic, and bronchitis. Take a look at what the FDA has to say about taking star anise and compare this to the claims made by companies selling it as a herbal remedy (http://www.fda.gov/ICECI/Enforcement Actions/EnforcementStory/EnforcementStoryArchive/ucm095929.htm). Could Donna's insistence on taking herbal remedies really do her any harm?

(Read more about this system in Daniel Chiras, *Human Biology, 7th edition*, Chapter 1.)

44. Medical Test Results (Contingent Probabilities)

Greta is finishing up her senior year of college and has decided to join the Peace Corps after graduation. In addition to filling out piles of application forms, she needed a complete physical exam. She was surprised to get a call from her doctor's office to come in to discuss some test results.

Dr. Bey informed Greta that her preliminary blood test indicated the possibility of a somewhat rare blood disorder that is known to occur in only one in a thousand individuals. The doctor pointed out that the test always finds the disease with 100% accuracy. However, the test is also associated with a 5% "false positive" rate; that is, of people without the disease, 5% nonetheless show a positive result on the test.

1. Greta and her doctor interpreted these facts to mean that she is almost certain to have the disease. However, if the disorder occurs in only one in a thousand individuals, then what do you think is the probability of Greta having the disease?

Greta isn't sure if she is more upset about her health or the fact that she won't be allowed into the Peace Corps. However, when you heard the facts in her case, you recalled one of your math professors talking about contingent probabilities. Despite Greta's diagnosis, you sit down with a pencil and paper to figure out the precise probability that she has the disease.

2. How many people have the disease out of 1000?

a. How many people do *not* have the disease out of 1000?

3. Given the false positive rate of 5%, how many *false* positives will there be out of those 999 people without the disease?

4. How many *true* positive results will there be out of 1000 people?

5. How many positive test results will there be (both true and false) out of 1000 people?

6. What percentage of all positive results are false positives?

a. Given that Greta has a positive test result, what is the probability that she has the disease?

45. Write Your Own Case Study

To enrich your understanding and application of anatomy and physiology, you will be asked to write your own case studies. These reports may be actual cases that you have encountered in your clinical experience or your personal life.

Part I: Begin with the patient's age and gender, and a brief history of the chief complaint that brought the patient to the hospital or to your attention. Note the important physical observations and assessments that were reported or observed when the patient entered your care or came to your attention. If you were involved with patient care, explain your role in the care of the patient. Write down your impressions of the patient's problems and the information you will need to fully understand the medical condition of this individual. If this is a friend, family member, or even yourself, give "the story" of the illness and all the details you know. Remember to exclude any details (such as the person's real name) that could be used to successfully identify the patient.

Part II: Focusing on the chief medical complaint and any interesting associated problem, discuss the underlying anatomy and physiology that explains the physical conditions of this patient. Also explain the physiology behind any tests and the physiology of how any treatments work.

Here is an example of how one of these case studies might be written up:

PART I

A six-year-old girl had a history of chronic ear infections. For years, she had been treated by her doctor with antibiotics with immediate success, but the ear infections continued to recur.

When the girl was four years old, her parents had noticed that she had difficulty hearing and were concerned about her language development. At this time, the girl was referred to an ear, nose, and throat specialist, who recommended that ear tubes be inserted into the girl's ears to prevent fluid buildup and consequent infections in the middle ear. The girl underwent minor surgery, and tubes were placed through the eardrums. After a course of antibiotics, the girl's ear infections cleared and she commented on how loud everything seemed.

A year later, while the ear tubes were still in place, the girl told her parents that she had "stuff coming out of her ears." She was immediately taken to the doctor and was put on a course of antibiotics. The diagnosis was that the girl had ear infections again despite the fact that the tubes were doing their job. Her ears continued to drain fluid, and she returned to the doctor. The doctor changed the antibiotic.

Several days later, the parents noticed that this rather active and robust six-year-old girl appeared thinner and was lethargic even though she had a ravenous appetite. She did not want to go to camp and complained of having a stomach ache. One evening after dinner, the girl threw up her entire meal. She was again taken to the doctor, and the doctor said that she was probably sensitive to the new antibiotics.

One day when the girl stayed home from camp, her mother noticed that she was going in and out of the house to go to the bathroom and to get a lot of water to drink. This pattern occurred throughout the day, and the mother became concerned that this behavior was unusual even on hot summer days. The girl's urine was tested for sugar. The test was positive, indicating diabetes.

PART II

1. The girl had chronic ear infections and her hearing was impaired. Why would her hearing be diminished? An ear infection most commonly occurs in the middle ear, which is connected to the throat via the Eustachian tube. If the middle ear is filled with fluid, the sound waves entering the auditory canal would strike the eardrum, but the eardrum would not move. For the person to hear, the vibrations of the sound waves must be changed to mechanical movement of the eardrum. The eardrum moves and transduces the vibrations to mechanical energy as the eardrum vibrates the three bones of the middle ear. The stapes bone is attached to the oval window, and again the movement is passed along to the cochlea. Hair cells in the cochlea send action potentials to the brain to be interpreted as sound of different pitch and loudness.

2. Fluid draining out of ear tubes is not normal. This sign indicated an ongoing infection, even though the girl had taken several courses of antibiotics. Which conditions would allow bacteria to continue to thrive? The doctor suspected that the infection was resistant to the prescribed antibiotic, but excess sugar is also an excellent environment for bacterial growth.

3. The child complained of a stomach ache. With diabetes, the body is starving, even though the individual consumes a lot of food. As part of a preservation mechanism, the body breaks down fat and muscle for energy instead of using the carbohydrates from food. The by-product of fat and muscle breakdown is a chemical called a ketone. Excess ketones in the body can cause vomiting and abdominal pain.

4. The child lost weight even though she had a good appetite. The food was not being used by her cells. The child was starving, despite the fact that she was eating a lot of food. The metabolic pathway to get glucose to her cells was not working. She lost weight because her cells could not absorb glucose.

5. The girl's urine was positive for glucose. Glucose is normally undetectable in the urine. Because of her lack of insulin, the glucose from digested food was not transported to her body cells. The excess glucose in her blood was not reabsorbed by the proximal convoluted tubule of the nephron and, therefore, was lost in the urine.

Chapter Opener Photograph Credits

Chapter 38: © S. Pearce/PhotoLink/Photodisc/Getty Images

Chapter 39: © Creatas

Chapter 40: © Andrea Weiss/ShutterStock, Inc.

Chapter 41: © AbleStock

Chapter 42: © Magdalena Bujak/ShutterStock, Inc.

Chapter 43: © Michel Cramer/ShutterStock, Inc.

Chapter 44: © Photodisc

Chapter 45: © Photos.com